Lecture Notes in Mathematics 1537

P. Fitzpatrick M. Martelli
J. Mawhin R. Nussbaum

Topological Methods for Ordinary Differential Equations

Lectures given at the 1st Session of the Centro Internazionale Matematico Estivo (C.I.M.E.) held in Montecatini Terme, Italy, June 24-July 2, 1991

Editors: M. Furi, P. Zecca

Springer-Verlag
Berlin Heidelberg New York
London Paris Tokyo
Hong Kong Barcelona
Budapest

Authors

Patrick Fitzpatrick
Department of Mathematics, University of Maryland
College Park, MD 20742, USA

Mario Martelli
Department of Mathematics, California State University
Fullerton, CA 92634, USA

Jean Mawhin
Institut Mathématique, Université de Louvain
B-1348 Louvain-La-Neuve, Belgium

Roger Nussbaum
Mathematics Department, Rutgers University
New Brunswick, NJ 08903, USA

Editors

Massimo Furi
Dipartimento di Matematica Applicata "G. Sansone"
Università di Firenze
Via S. Marta 3, I-50139 Firenze, Italy

Pietro Zecca
Dipartimento di Sistemi e Informatica
Università di Firenze
Via S. Marta 3, I-50139 Firenze, Italy

Mathematics Subject Classification (1991): 34A02, 34B15, 47H10, 54H25

ISBN 3-540-56461-6 Springer-Verlag Berlin Heidelberg New York
ISBN 0-387-56461-6 Springer-Verlag New York Berlin Heidelberg

Printed in Germany

Typesetting: Camera-ready by author/editor
46/3140-543210 - Printed on acid-free paper

PREFACE

The topological degree is a fundamental tool for proving the existence of various kinds of solutions of nonlinear differential equations and for investigating the structure of such sets of solutions. Since the original classical work of Leray and Schauder, many authors have made contributions to the problem of extending the Leray-Schauder degree and applying it to new problems in analysis. These generalizations range from extensions of the Lefschetz fixed point theorem and the fixed point index on ANR's (manifolds and finite unions of convex sets providing important examples of ANR's) to the theory of parity of one parameter families of Fredholm operators, and from the theory of coincidence degree for mappings on Banach spaces to homotopy methods for continuation principles.

The motivation for a CIME session on these topics arose from the observation that very few of the CIME sessions which have been held over the years have been devoted to arguments related to Topological Methods in Analysis. We mention a session on Nonlinear Differential Equations (1964), one on Problems on Nonlinear Analysis (1970) and one on Bifurcation Theory and Applications (1983). However, since none of these previous sessions was entirely devoted to the growing field of topological methods in the theory of ordinary differential equations, the intention of this CIME session was to present the state of the art (at least for certain topological methods) and to provide a forum for discussion of the wide variety of mathematical tools which are involved.

Five CIME courses were given by well-known, active mathematicians with extensive experience in the application of topological methods to boundary value problems for ordinary differential equations. The texts for four of these courses are contained in this volume.

We are proud to have organized this CIME session, and we are grateful to the lecturers for their efforts at lucid exposition. We thank the Director and the staff of CIME for their support.

Massimo Furi, Pietro Zecca

TABLE OF CONTENTS

THE PARITY AS AN INVARIANT FOR DETECTING BIFURCATION OF THE ZEROES OF ONE PARAMETER FAMILIES OF NONLINEAR FREDHOLM MAPS*

PATRICK FITZPATRICK
DEPARTMENT OF MATHEMATICS
UNIVERSITY OF MARYLAND
COLLEGE PARK, MD., U.S.A.

CONTENTS

INTRODUCTION

The Rabinowitz Global Bifurcation Theorem [**31**] improves an old bifurcation result of Krasnosel'skii [**22**] by obtaining conclusions in the large about bifurcating branches of nontrivial zeroes of a family of compact vector fields parametrized by an interval. A few years ago, Jacobo Pejsachowicz and I became interested in finding the appropriate extension of this global bifurcation theorem to one parameter families of nonlinear elliptic differential operators subject to general boundary conditions of Shapiro-Lopatinskij type. Generally, the maps induced in function spaces by this type of operator are nonlinear Fredholm. They do not possess any obvious reduction to compact vector fields. In trying to extend the global bifurcation theorem to this class of problems, a difficulty arises: the standard proof using the homotopy invariance of the Leray-Schauder degree cannot possible work any more, since no degree theory which extends the Leray-Schauder degree to a class of mappings which includes all linear isomorphisms can be homotopy invariant.

* **Supported by a NATO Research Grant.**

Of course, there have been a number of different extensions of the Leray-Schauder degree to a homotopy invariant degree. Often, these involve choices of classes of nonlinear Fredholm maps whose topology resembles that of the class on compact vector fields. In [**10,15**], Pejsachowicz and I defined an additive, integer valued topological degree for the class of quasilinear Fredholm mappings, a class introduced by Snirel'man [**32**] in his study of the nonlinear Riemann-Hilbert problem. General fully nonlinear elliptic boundary value problems can be formulated as the zeroes of quasilinear Fredholm mappings (cf. [**1**]). This class includes all linear isomorphisms, so the degree could not be homotopy invariant. However, to each path of linear Fredholm operators with invertible ends we assigned a homotopy invariant which we called the *parity.* The homotopy property of our degree could be perfectly described in terms of the parity. This enabled us to find the appropriate extension of the global bifurcation theorem for families parametrized by an interval. Moreover, for parameter spaces with nontrivial topology it revealed bifurcations for the zeroes of families of nonlinear Fredholm mappings which are forced by the topological nontriviality of the parameter space and which have no correspondent for families of compact vector fields.

In fact, the parity plays an important part in the study of multiplicity and bifurcation for the zeroes of families of general nonlinear Fredholm mappings parametrized by general spaces (cf. [**29**], [**13**], [**8**]). In these lectures, I will describe some of the recent results [**10,11,12,13,14,15**] regarding the role of the parity in the study of local and global bifurcation of the zeroes of one parameter families of nonlinear Fredholm maps. By first placing the classical bifurcation results in a way which is a little more geometric than the way they are usually described, my aim is to show that the parity appears quite naturally as a topological invariant of the family of linearizations about a trivial branch of zeroes whose nontriviality forces bifurcation for the zeroes of the associated family of nonlinear operators.

In the first section, I consider the set $L(R^n)$ of linear operators acting in R^n, describing the two components of the group $GL(R^n)$ of invertible operators in $L(R^n)$, the stratification of the singular operators $\mathcal{S}(R^n) = L(R^n) \setminus GL(R^n)$ and the formula which relates the number of times that a generic path in $L(R^n)$ having invertible ends crosses the singular set with the position of the end points of the path. This is preparation for the second section in which I begin with a similar description for the set of all linear compact vector fields acting on a real Banach space. This, in turn, is the basis for an understanding of certain topological properties of $\Phi_0(X,Y)$, the set of all linear Fredholm operators of index zero acting between the real Banach spaces X and Y, which are relevant to bifurcation problems and degree theory. As in the finite dimensional case, in $\Phi_0(X,Y)$ one singles out the set of invertible operators $GL(X,Y)$ and its complement, the set of singular operators $\mathcal{S}(X,Y)$. To each path $L: I \to \Phi_0(X,Y)$ with invertible ends there is assigned its parity $\sigma(L,I) \in \{\pm 1\}$. The parity is a homotopy invariant of such paths. In fact, the parity of a path is 1 if and only if the path can be deformed out of the set of singular operators. Generically, the parity is an intersection index, the parity being 1 if and only if the path intersects the singular set an even number of times. The parity provides a simple description of the Poincaré group of $\Phi_0(X,Y)$. This is particularly interesting since $\Phi_0(X,Y)$ may be connected. According to Kuiper's Theorem, if X is a Hilbert space, then $GL(X,Y)$ is contractible and this implies that

$\Phi_0(X,Y)$ is connected. But it is not simply-connected, and the parity of a closed path defines a homeomorphism of the Poincaré group of $\Phi_0(X,Y)$ with Z_2.

In §3, I turn to a discussion of the significance of the parity in the study of local bifurcation for the zeroes of a one parameter families of nonlinear Fredholm mappings. Assume such a family has a branch of trivial solutions. Linearizing the family at the trivial branch one gets a path of linear Fredholm operators. At an isolated singular point of this path, the transversal crossing of $\mathcal{S}(X,Y)$ is shown to be a necessary and sufficient condition on the linearization in order that there bifurcate a curve of nontrivial solutions (cf. **[18]**, [3], **[21]**). That the parity of the restriction of the path to a small neighborhood of an isolated singular point be -1 is shown to be a necessary and sufficient condition for bifurcation (cf. **[22]**, **[6]**, **[20]**). Methods of computing the parity at points where a path crosses $\mathcal{S}(X,Y)$ degenerately are also briefly described as specific conditions that ensure bifurcation (cf. **[34]**, **[33]**).

To show that the properties of the path of linearizations which imply local bifurcation actually imply bifurcation in the large, it is necessary to work in a class of mappings for which there is a degree theory in which the homotopy dependence of the degree is effectively described. In §4, I describe the construction of the degree for quasilinear Fredholm mappings and some of its properties, including special cases of the homotopy property and the regular value formula (**[10,15]**).

In the last section, this degree is first used to establish the natural generalization of the Rabinowitz Global Bifurcation Theorem for families of quasilinear Fredholm mappings parametrized by an interval. Then the degree is used to prove a result on bifurcation to infinity for a family of quasilinear Fredholm maps parametrized by the circle. As I already mentioned, there are paths $L: S^1 \to \Phi_0(X,Y)$ which cannot be deformed in $\Phi_0(X,Y)$ to constant paths. These are precisely the closed paths which have parity -1, and the perturbation of such a path by a closed path of compact operators also has the same property. For paths of differential operators this means that this property is determined by the "top-order terms". In the semilinear case, the last result I describe asserts that if $f: S^1 \times X \to Y$ has the form $f(\lambda, x) = L_\lambda x + C(\lambda, x)$, where $L: S^1 \to \Phi_0(X,Y)$ has parity -1 and $C: S^1 \times X \to Y$ is compact, then if there is one parameter value λ_0 at which the degree of f_{λ_0} on X is nonzero, then there is an unbounded branch of solutions of the equation $f(\lambda, x) = 0$ **[13]**. An unusual feature of this result is that it ensures "bifurcation to infinity" without imposing any assumptions at infinity.

§1 The Determinant and the Structure of $L(R^n)$

Let V be a finite dimensional vector space, either over R or C. Let $L(V)$ denote the space of linear operators from V into itself, and $GL(V)$ be the group of invertible operators in $L(V)$. If the scalar λ is an eigenvalue of $L \in L(V)$ and n is the dimension of V, then Null $(\lambda I - L)^n \oplus$ Im $(\lambda I - L)^n = V$ and each of the summands is invariant with respect to L. The nonzero members of Null $(\lambda I - L)^n$ are called generalized eigenvectors of L corresponding to the eigenvalue λ and the dimension of this space, $m(L, \lambda)$, is called the algebraic multiplicity of λ as an eigenvalue of L. The restriction of L to Im $(\lambda I - L)^n$

does not have λ as an eigenvalue while the only eigenvalue of the restriction of L to Null $(\lambda I - L)^n$ is λ. From this follows the following classical result, a proof of which I include since it is also the basis of an interesting generalization.

Proposition 1.1 *If L is in $GL(R^n)$, define $\epsilon(L) = (-1)^m$, where m is the sum of the algebraic multiplicities of the negative eigenvalues of L. Then $GL(R^n)$ has two connected components which are labeled by the values of the function ϵ.*

Proof. Define $E_-(L)$ to be the sum of the generalized eigenspaces corresponding to the negative eigenvalues of L. The preceding comments imply that there is a subspace $E_+(L)$ of R^n which is invariant under L, the restriction of L to which has no negative eigenvalues and for which one has the decomposition $R^n = E_-(L) \oplus E_+(L)$. Define $S: [0,1] \to L(R^n)$ by

$$S_t|_{E_+} = tL|_{E_+} + (1-t)Id|_{E_+} \text{ and } S_t|_{E_-} = tL|_{E_-} - (1-t)Id|_{E_-}.$$

Since the eigenvalues of the restriction of L to $E_-(L)$ consist exactly of the $\lambda_i's$ and the restriction of L to $E_+(L)$ has no negative eigenvalues, each S_t is invertible. Thus L has been deformed in $GL(R^n)$ to an operator T for which there is a decomposition $R^n = E_- \oplus E_+$ with respect to which T is the identity on E_+ and minus the identity on E_-.

Now, on a two dimensional space, the identity operator may be deformed through a path of rotations to minus the identity. Thus, if E_- has even dimension, we may deform T into the identity. On the other hand, if E_- has odd dimension we may deform T to an operator which is minus the identity on a one dimensional subspace and the identity on a complementary subspace. Moreover, it is clear that any two linear operators of this latter type may be joined by a path of linear isomorphisms.

So $GL(R^n)$ has at most two components. It remains to show that it has precisely two components which are labeled by the function ϵ. This amounts to showing that ϵ is continuous. To do this, choose any basis for R^n and for $T \in GL(R^n)$ let det (T) be the determinant of the matrix which represents T with respect to this basis. But $\det(T)$ varies continuously with T, and hence so does $\epsilon(T) = \text{sgndet } T$. **Q. E. D.**

The sign of the determinant characterizes the parity of the sum of the negative eigenvalues of T. To foreshadow an extension of the preceding proposition to the infinite dimensional case, we now characterize the parity of the negative eigenvalues of L in terms of the dimension of a spectral projection.

Let A be an $n \times n$ matrix of complex numbers. Define the characteristic polynomial $p(z)$ by $p(z) = \det[z - A]$. Choose a simple closed curve Γ on which no eigenvalues of A lie. Define $m(A, \Gamma)$ to be the number of eigenvalues of A, counted with algebraic multiplicity, which are interior to Γ. The multiplicity of an eigenvalue coincides with the order of the eigenvalue as a root of $p(z)$, so that, by the Logarithmic Residue Formula for the zeroes of the analytic function, we have the following integral representation of $m(A, \Gamma)$:

$$m(A, \Gamma) = \frac{1}{2\pi i} \oint_\Gamma \frac{p'(z)}{p(z)} \, dz. \tag{1.2}$$

But recall the following formula for the derivative of the determinant of $z - A$:

$$\frac{d}{dz} \det(z - A) = \det(z - A) \cdot \text{trace } [z - A]^{-1}.$$

Substituting this in equation (1.2) and using the linearity of the trace, we obtain

$$m(A,\Gamma) = \text{trace}\,[\frac{1}{2\pi i}\oint_\Gamma (z-A)^{-1}\,dz]. \tag{1.3}$$

Define

$$P(A,\Gamma) \equiv \frac{1}{2\pi i}\oint_\Gamma (z-A)^{-1}\,dz. \tag{1.4}$$

Using the Cauchy Integral Formula and the resolvent identity *, it follows that $P(A,\Gamma)$ is a projection. However, the trace of a projection is just the dimension of its range. Hence, using (1.3) we arrive at the formula

$$m(A,\Gamma) = \dim \operatorname{Im} P(A,\Gamma) \qquad (1.5).$$

Of course, fixing a basis in the n-dimensional space V, we may identify $L(V)$ with the set $M_{n\times n}(R)$ of $n \times n$ real matrices, if V is real, and the set $M_{n\times n}(C)$ of $n \times n$ complex matrices, if V is complex. By this identification, we may consider the concepts of determinant and trace to be defined either for members of $L(V)$ or for the corresponding set of $n \times n$ matrices. Moreover, if $L \in L(R^n)$ and Γ is a curve in the complex plane which contains no eigenvalues of L, then $P(L,\Gamma)$ and $m(L,\Gamma)$ are properly defined.

Let $L \in L(R^n)$ be invertible. Choose a curve Γ in the complex plane, which is symmetric with respect to the real axis, on which lie no eigenvalues of L and which encloses all of the negative real eigenvalues of L but no other real eigenvalues. Define $p(z) = \det[z - L]$ for $z \in C$. Since $p(z)$ is a polynomial with real coefficients and Γ was chosen to be symmetric with respect to the real axis and contain no positive real zeroes in its interior, the roots of $p(z)$ inside Γ which are not both real and negative appear in conjugate pairs of equal algebraic multiplicities, and hence $\epsilon(L) = (-1)^{m(L,\Gamma)}$. Thus

$$\epsilon(L) = (-1)^{m(L,\Gamma)} \text{ where } m(L,\Gamma) = \dim \operatorname{Im} P(L,\Gamma). \tag{1.6}$$

Formula (1.4) defines a projection which is called the *spectral projection for A inside* Γ. The integer $m(A,\Gamma)$ is called the *spectral multiplicity of A inside* Γ. These concepts have infinite dimensional correspondents which we will describe in the next section.

I will now turn to a description of the complement of the group of invertible operators, namely, the set of singular linear operators on R^n, $\mathcal{S}(R^n) = L(R^n) \setminus GL(R^n)$. This set may be decomposed as

$$\mathcal{S}(R^n) = \bigcup_{k=i,\ldots,n} \mathcal{S}_k(R^n),$$

---

* The resolvent identity, sometimes called the Hilbert identity, simply asserts that if the scalars λ and μ are not in the spectrum of $L \in L(X)$, then

$$(\lambda - L)^{-1} - (\mu - L)^{-1} = (\mu - \lambda)(\lambda - L)^{-1} \circ (\mu - L)^{-1}.$$

where each $\mathcal{S}_k(R^n)$ consists of those operators with a k-dimensional nullspace. It is convenient to say that a C^1 path $L: R \to L(R^n)$ has property (T) at λ_0 if

$$\dim \text{Null } L_{\lambda_0} = 1$$

and $\qquad\qquad\qquad\qquad (T)$

$$L'_{\lambda_0}(\text{Null } (L_{\lambda_0})) \oplus \text{Im } (L_{\lambda_0}) = R^n.$$

Proposition 1.7 *The stratum $\mathcal{S}_1(R^n)$ of $\mathcal{S}(R^n)$ is a codimension 1 submanifold of $L(R^n)$. Moreover, for a C^1 path $L: R \to L(R^n)$, the following three properties are equivalent:*

$$\textit{The path } L: R \to L(R^n) \textit{ crosses } \mathcal{S}_1(R^n) \textit{ transversally at } \lambda_0. \tag{i}$$

$$\det L_{\lambda_0} = 0 \textit{ and } (d/d\lambda) \det L_\lambda|_{\lambda=\lambda_0} \neq 0. \tag{ii}$$

$$\textit{The path } L: R \to L(R^n) \textit{ has property (T) at } \lambda_0. \tag{iii}$$

Proof. Of course, $L \in \mathcal{S}(R^n)$ if and only if $\det L = 0$ and one also observes that if $L \in \mathcal{S}(R^n)$, then $L \in \mathcal{S}_1(R^n)$ if and only if $\text{cof } L \neq 0$. So

$$\mathcal{S}_1(R^n) = \{L \in L(R^n) | \det L = 0 \text{ and cof } L \neq 0\}.$$

On the other hand, if $A, B \in M_{n\times n}(R)$, we have the following classical formula, which is a consequence of linearity and its obvious validity in the case that B is elementary:

$$< \nabla \det(A), B > \equiv \lim_{t\to 0} \frac{\det[A + tB] - \det[A]}{t} = \text{trace } [\text{cof A} \cdot B]. \tag{1.8}$$

Thus,

$$\mathcal{S}_1(R^n) = \{L \in L(R^n) | \det L = 0 \text{ and } \nabla \det L \neq 0\}. \tag{1.9}$$

This description of $\mathcal{S}_1(R^n)$, together with the Implicit Function Theorem, implies that $\mathcal{S}_1(R^n)$ is a codimension 1 submanifold of $L(R^n)$.

Now consider a C^1 path $L: R \to L(R^n)$ and $\lambda_0 \in R$. In view of (1.9), this path is transverse to $\mathcal{S}_1(R^n)$ at λ_0 if and only if $\det L_{\lambda_0} = 0$ and $(d/d\lambda) \det(L_\lambda)|_{\lambda=\lambda_0} \neq 0$. But (1.8) implies that

$$\frac{d}{d\lambda} \det(L_\lambda)|_{\lambda=\lambda_0} = \text{trace } [\text{cof } L_{\lambda_0} \cdot L'_{\lambda_0}],$$

so that $L: R \to L(R^n)$ is transverse to $\mathcal{S}_1(R^n)$ at λ_0 if and only if

$$\det L_{\lambda_0} = 0 \text{ and trace } [\text{cof } L_{\lambda_0} \cdot L'_{\lambda_0}] \neq 0.$$

The equivalence of transversal crossing and property (T) now follows from the observation that if $A, B \in M_{n\times n}(R)$, then $\text{trace } [\text{cof } A \cdot B] \neq 0$ if and only if $B(\text{Null } A) \oplus \text{Im } A = R^n$. **Q.E.D.**

There are two special cases in which the transversal crossing of $\mathcal{S}_1(R^n)$ can be precisely related to the behavior of the spectrum of L_λ as a function of λ near λ_0. The first very special case is when the path has the form $L_\lambda = \lambda I - A$. A singular point of L is an eigenvalue of A. As usual, an eigenvalue λ of the matrix A is defined to be simple provided that dim Null $(\lambda I - A) = 1$ and Null $(\lambda I - A) =$ Null $(\lambda I - A)^2$. One easily checks to see that an eigenvalue λ of A is simple if and only if $d/d\lambda \det(\lambda I - A)|_{\lambda=\lambda_0} \neq 0$. So, in view of the equivalence of (i) with (ii) in the preceding proposition, we obtain the first of the following two results.

Proposition 1.10 *Let $A \in L(R^n)$. Define $L: R \to L(R^n)$ by $L_\lambda = \lambda I - A$. Then $L: R \to L(R^n)$ crosses $\mathcal{S}_1(R^n)$ transversally at λ_0 if and only if λ_0 is a simple eigenvalue of A.*

Proposition 1.11 *Let $L: R \to L(R^n)$ be C^1 and suppose that $\lambda_0 \in R$ has the property that 0 is a simple eigenvalue of L_{λ_0}. Then $L: R \to L(R^n)$ crosses $\mathcal{S}_1(R^n)$ transversally at λ_0 if and only if $\epsilon'(\lambda_0) \neq 0$, where I is a neighborhood of λ_0 and $\epsilon: I \to R$ is the unique C^1 path with $\epsilon(\lambda_0) = 0$ and $\epsilon(\mu)$ an eigenvalue of L_μ for each μ in I.*

Proof. Define $h: R \times R \to R$ by $h(\lambda, \mu) = \det[L_\lambda - \mu I]$. The assumption that 0 is a simple eigenvalue of L_{λ_0} means exactly that $\partial h/\partial\mu(\lambda_0, 0) \neq 0$. Thus, by the Implicit Function Theorem, the zeroes of h in a neighborhood of $(\lambda_0, 0)$ comprise the graph of a C^1 function $\epsilon: I \to R$ with $\epsilon(\lambda_0) = 0$. Since $h(\lambda, \epsilon(\lambda)) = 0$ for all $\lambda \in I$, upon differentiating we get

$$\frac{\partial h}{\partial \lambda}(\lambda_0, 0) + \frac{\partial h}{\partial \mu}(\lambda_0, 0) \cdot \epsilon'(\lambda_0) = 0.$$

But since $\partial h/\partial\mu(\lambda_0, 0) \neq 0$, it follows that $d/d\lambda(\det L_\lambda)|_{\lambda=\lambda_0} \equiv \partial h/\partial\lambda(\lambda_0, 0) \neq 0$ if and only if $\epsilon'(\lambda_0) \neq 0$. The result now follows from Proposition 1.7. **Q.E.D.**

We conclude this section with a description of the strata $\mathcal{S}_k(R^n)$ when $k > 1$. In fact, $\mathcal{S}_k(R^n)$ is a submanifold of $L(R^n)$ which is of codimension k^2 in $L(R^n)$. To see this, fix $L \in \mathcal{S}_k(R^n)$. By composing with an element of $GL(R^n)$, we may suppose that L=P is a projection onto the subspace V of R^n of dimension $n-k$. Denote the range of $I-P$ by W. Let $T \in L(R^n)$ with $||T|| < 1$. Then $I+T \in GL(R^n)$. Now $P+T \in \mathcal{S}_k(R^n)$ if and only if Im $(P+T)$ has dimension $n-k$. Since $(P+T)V = (I+T)V$ has dimension $n-k$, we see that $P+T \in \mathcal{S}_k(R^n)$ if and only if $T(W) \subset (I+T)V$, or, equivalently, $(I+T)^{-1} \circ T$ maps W into V. Choose $w_1, \ldots, w_k$ to be a basis of W, and let $w'_1, \ldots, w'_k$ be a basis for the orthogonal complement of V. Then the functions $\psi_{i,j}: L(R^n) \to R$, defined for $1 \leq i, j \leq k$ by $\psi_{i,j}(T) = < (I+T)^{-1}Tw_i, w'_j >$, are the components of the mapping $\Psi: L(R^n) \to R^{k^2}$ which has the property that, in a ball of radius 1 about P, the zeroes of the map $P+T \mapsto \psi(T)$ coincide with the members of $\mathcal{S}_k(R^n)$. It is easy to see that the derivative, at P, of this map is surjective, and so, by the Implicit Function Theorem, $\mathcal{S}_k(R^n)$ is a submanifold of $L(R^n)$ which is of codimension k^2 in $L(R^n)$.

It now follows from standard transversality results [**17**] that a path in $L(R^n)$ with invertible ends may be perturbed so that the perturbation has only a finite number of singular points at each of which there is a transversal crossing of $\mathcal{S}_1(R^n)$. But, by the equivalence of (i) and (ii) in Proposition 1.7, the determinant changes sign each time a path crosses $\mathcal{S}_1(R^n)$ transversally. Thus, in view of Proposition 1.1, the parity of the number of intersections of the perturbation is even if and only if the ends of the original path lie in the same component of $GL(R^n)$, assuming the original path has invertible ends. In summary, we have

Theorem 1.12 *The group of invertible operators $GL(R^n)$ has two connected components, which are labeled by the function sgndet. The set of singular operators may be stratified by $\mathcal{S}(R^n) = \bigcup_{k=i,\dots,n} \mathcal{S}_k(R^n)$, where each $\mathcal{S}_k(R^n)$ is a submanifold of $L(R^n)$ of codimension k^2. Moreover, a C^1 path $L: R \to L(R^n)$ crosses $\mathcal{S}(R^n)$ transversally at λ_0 if and only the path satisfies property (T) at λ_0. Finally, given a continuous path $L: I = [a, b] \to L(R^n)$ with invertible ends, there are arbitrarily close smooth perturbations of L which cross $\mathcal{S}(R^n)$ transversally and*

$$sgndet\ L_a \cdot sgndet\ L_b = (-1)^m, \tag{1.13}$$

where m is the number of singular points of any such small perturbation.

§2 The Parity and the Structure of $\Phi_0(X, Y)$

For many real Banach spaces X, including infinite dimensional Hilbert spaces, the general linear group, $GL(X)$, is contractible. For such a space, $GL(X)$ is connected, so there is no correspondent of Proposition 1.1, and, moreover, the set of linear Fredholm operators of index 0, $\Phi_0(X, Y)$, is also connected. But $\Phi_0(X, Y)$ is not simply connected. I shall now describe a homotopy invariant of paths of linear Fredholm operators with invertible ends called the parity. The parity was introduced in [**15**]. It is nontrivial if and only if the path can be deformed, through paths with invertible ends, into $GL(X, Y)$. Generically, it is an intersection index which counts, mod 2, the number of singular points of the path. Roughly speaking, while each of the terms in the product in the left-hand side of formula (1.13) has lost its significance, it is possible to assign a meaning, the parity, to the right-hand side of (1.13). In the succeeding sections, the relevance of the parity to degree theory and bifurcation problems will be described. This present section will be devoted to a description of the parity function itself. It is useful to first describe spectral projections and then consider the special subset of $\Phi_0(X)$ consisting of the linear compact vector fields.

Let X and Y be Banach spaces, either over R or C. By $L(X, Y)$ we denote the space of bounded linear operators from X to Y, and consider $L(X, Y)$ as a Banach space with the usual norm. By $GL(X, Y)$ we denote the set of invertible operators in $L(X, Y)$. An operator T in $L(X, Y)$ is called Fredholm if the nullspace of T, Null T, has finite dimension and the image of T, Im T, is of finite codimension in Y. We denote the set of Fredholm operators by $\Phi(X, Y)$. For $T \in \Phi(X, Y)$, the numerical Fredholm index of T, Ind(T), is defined by Ind $(T) = \dim(\text{Null } T) - \text{codim Im } T)$. The subset of $L(X, Y)$ consisting of Fredholm operators of index m is denoted by $\Phi_m(X, Y)$.

The spectrum of $L \in L(X)$ consists of those scalars λ such that $\lambda I - L \notin GL(X)$. Those scalars not in the spectrum of L comprise the resolvent of L. Points λ in the spectrum which have property that Null $\lambda I - L$ is nontrivial are called eigenvalues.

Definition *An eigenvalue λ of $L \in L(X)$ is said to be of finite algebraic multiplicity if $\lambda I - L$ is Fredholm of index 0 and Null $(\lambda I - L)^m =$ Null $(\lambda I - L)^{m+1}$ for some* *integer m. The algebraic multiplicity of λ as an eigenvalue of L is the dimension of Null $(\lambda I - L)^m$. An eigenvalue of algebraic multiplicity 1 is said to be simple.*

Using a characterization of linear Fredholm operators which I will shortly describe, it is easy to see that an eigenvalue λ of $L \in L(X,Y)$ is of finite algebraic multiplicity if and only if there is an integer m such that Null $(\lambda I - L)^m$ has finite dimension and

$$\text{Null } (\lambda I - L)^m \oplus \text{ Im } (\lambda I - L)^m = X.$$

Moreover, λ is a simple eigenvalue of L if and only if dim Null $\lambda I - L$ has dimension 1 and

$$\text{Null } (\lambda I - L) \oplus \text{ Im } (\lambda I - L) = X.$$

Recall that an operator $L \in L(X)$ is called a linear compact vector field if it is a compact perturbation of the identity. The set of invertible compact vector fields in $L(X)$ forms a group which we will denote by $GL_C(X)$.

The Riesz-Schauder Theorem * *Let X be a Banach space and let $L \in L(X)$ be a compact vector field. Then L is Fredholm of index 0. In fact, if the scalar $\lambda \neq 1$ is in the spectrum of L, then λ is an eigenvalue of finite algebraic multiplicity.*

Using the fact that $GL(X)$ is open in $L(X)$, it follows from the Riesz-Schauder Theorem that is that if $T \in L(X)$ is a compact vector field, then the points in its spectrum, not equal to 1, are eigenvalues which are isolated points of its spectrum. Consequently, a linear compact vector field has only a finite number of negative eigenvalues, each of which is of finite algebraic multiplicity.

Proposition 1.1 asserts that the general linear group $GL(R^n)$ has two connected components. If X is a real Banach space, so also does the group $GL_C(X)$ of invertible linear compact vector fields. In order to verify this, it is useful to describe the infinite dimensional correspondent of (1.4). This is the subject of another classical result.

The Riesz Projection Theorem ** *Let X a complex Banach space and $L \in L(X)$. Suppose that the simple, closed curve $\Gamma \subset C$ lies in the resolvent of L. Then*

$$P \equiv P(L, \Gamma) = \frac{1}{2\pi i} \oint_\Gamma (z - L)^{-1} \, dz$$

defines a projection. The corresponding decomposition $X = Im\ P \oplus Im\ (I - P)$ reduces L into operators $L_P \oplus L_{I-P}$ with respect to which the spectrum of L, sp(L), decomposes as

$$sp(L_P) = sp(L) \cap int\Gamma \text{ and } sp(L_{I-P}) = sp(L) \cap ext\Gamma.$$

* Chap.1 of **[16]**
** Chap.2 of **[16]**

Given X a complex Banach space and L and Γ as in the Riesz Projection Theorem, we call the operator $P(L,\Gamma)$ the *spectral projection for L inside* Γ and define the *spectral multiplicity of L inside* Γ, $m(L,\Gamma)$, by the formula

$$m(L,\Gamma) \equiv \dim \operatorname{Im} P(L,\Gamma).$$

From the finite dimensional case and the Riesz Projection Theorem, it is not difficult to see that $m(L,\Gamma)$ is finite if and only if the spectrum of L which is interior to Γ comprises a finite number of eigenvalues, each of finite algebraic multiplicity. In this case, it follows from the finite dimensional formula (1.13) that $m(L,\Gamma)$ equals the sum of the algebraic multiplicities of the eigenvalues of L inside of γ.

In general, if P and Q are any projections in a Banach space, then

$$[(I-P)+Q]\circ P = Q\circ P = Q\circ[(I-Q)+P].$$

Thus, if $||P-Q|| < 1$, then $I \pm (P-Q)$ are isomorphisms and so the images of P and Q have the same dimension. Also, inversion is continuous among invertible linear operators acting between Banach spaces. From these two observations we obtain the following

Corollary 2.1 *Let X be a complex Banach space and consider a continuous path $L\colon I \to L(X)$. Suppose that Γ is a simple, closed curve in the complex plane which contains none of the spectra of $L_t, t \in I$. Then $m(L_t,\Gamma)$ is independent of t in I.*

In order to use the Riesz Projection Theorem and its corollary to study operators which act between real, rather than complex, Banach spaces, one can complexify the space and the operator. Specifically, given a Banach space X over R, there is an obvious complex structure on $X \oplus X$ making it a complex Banach space. We denote this space by X_C. Then every $T \in L(X)$ has a unique complex-linear extension to $T_C \in L(X_C)$. The complex-linear extension of a compact vector field is again a compact vector field.

Proposition 2.2 *Let X be a real Banach space. For $T \in GL_C(X)$, define*

$$\epsilon(T) = (-1)^m,$$

where m is the sum of the algebraic multiplicities of the negative eigenvalues of T. Then $GL_C(X)$ has two connected components which are labeled by the values of the function ϵ.

Proof. By using the convexity of the class of linear compact vector fields and the Riesz-Schauder Theorem one can follow the proof of the first part of Proposition 1.1 and conclude that the subsets of $GL_C(X)$ on which ϵ is constant are connected.

To verify that $GL_C(X)$ has exactly two components which are determined by the values of the function ϵ now amounts to showing that ϵ is continuous.

Let $T \in GL_C(X)$. We complexify to obtain $T_C \in GL_C(X_C)$. Since the negative real spectrum of T_C consists of points which are isolated in the spectrum of T_C, we may choose Γ to be a simple, closed curve in the left-side of the complex plane which is symmetric with respect to the real axis, which lies in the resolvent of T_C and for which the real spectrum of T_C in its interior consists exactly of the set of all the real negative eigenvalues of T. Since inversion is continuous among linear bounded operators in a

Banach space, we may choose a neighborhood $\mathcal{N}$ of T, in $L(X)$, having the property that if $S \in \mathcal{N}$, then the spectrum of the segment joining T_C and S_C does not intersect Γ and the real negative spectrum of S_C comprises the real spectrum of S_C that lies entirely in the interior of Γ. By Corollary 2.1, $m(S_C, \Gamma)$ is independent of the choice of $S \in \mathcal{N}$. On the other hand, if $S \in \mathcal{N}$, then inside of Γ the eigenvalues of S_C occur in conjugate pairs of equal finite algebraic multiplicity, so that $\epsilon(S) = (-1)^{m(S_C,\Gamma)}$. Thus ϵ is also constant on $\mathcal{N}$. **Q.E.D.**

Corollary 2.3 *Let X be a real Banach space. If $L, T \in GL_C(X)$, then $\epsilon(L \circ T) = \epsilon(L) \cdot \epsilon(T)$. In particular, $\epsilon(L) = \epsilon(L^{-1})$. Moreover, if $S \in GL(X,Y)$, then $\epsilon(T) = \epsilon(S \circ T \circ S^{-1})$.*

Proof. Assume $\epsilon(L) = 1$. Then L and the identity can be joined by a path γ in $GL_C(X)$. Then it is clear that $t \mapsto \gamma(t) \circ T$ and $t \mapsto S \circ \gamma(t) \circ S^{-1}$ define paths in $GL_C(X)$ joining the T to $L \circ T$ and $S \circ T \circ S^{-1}$, respectively. From the continuity of ϵ, it follows that $\epsilon(L \circ T) = 1$ and $\epsilon(S \circ T \circ S^{-1}) = 1$. The case when $\epsilon(L) = -1$ is similar. **Q.E.D.**

We now turn to a discussion of the space of linear Fredholm operators (cf. **[16]**, **[35]**).

Proposition 2.4 *Let $L \in L(X,Y)$. Then the following three assertions are equivalent:*
(i) The operator L is Fredholm of index 0.
(ii) There is a compact operator $K \in L(X,Y)$ such that $L + K$ is invertible.
(iii) There is an isomorphism $S \in L(Y,X)$ such that $S \circ L$ is a compact vector field.

Proof. Let $L \in \Phi_0(X,Y)$. Since Null L is finite dimensional, there is a continuous projection P of X onto Null L. Choose a closed subspace V of Y which is a linear complement in Y of Range L. By definition, Null L and V are of the same finite dimension, so we may select $S \in GL(\text{Null } L, V)$. Define $K = S \circ P$, and observe that K is compact and $L + K$ is an isomorphism. Thus (i) implies (ii).

To verify that (ii) implies (iii), choose K compact so that $L + K$ is invertible, define $S = (L+K)^{-1}$, and observe that $S \circ L = I - (L+K)^{-1} \circ K$.

Finally, that (iii) implies (i) is a consequence of the the assertion in the Riesz-Schauder Theorem that linear compact vector fields are Fredholm of index 0 and of the observation that the composition of a Fredholm operator of index 0 with an isomorphism is again Fredholm of index 0. **Q.E.D.**

For $L \in \Phi_0(X,Y)$, an operator $S \in GL(Y,X)$ having the property that $S \circ L$ is a compact vector field will be called a *parametrix* for L. It is clear that if S and $\bar{S}$ are each parametrices for L, then $\bar{S}^{-1} \circ S \in GL_C(Y)$.

Our primary interest is in parametrized families of operators. Given a topological space Λ, by a *family of linear Fredholm operators parametrized by* Λ we mean a continuous mapping $L: \Lambda \to \Phi_0(X,Y)$. A *parametrix* for a family $L: \Lambda \to \Phi_0(X,Y)$ is a family of invertible operators $S: \Lambda \to GL(Y,X)$ having the property that each $S(\lambda) \circ L(\lambda)$ is a linear compact vector field.

It turns out that families of Fredholm operators parametrized by contractible metric spaces always have parametrices.

Theorem 2.5 (cf. [35], [15], [23]) *Every family of linear Fredholm operators parametrized by a contractible paracompact space has a parametrix.*

Given $I = [a,b] \subset R$, a family $L: I \to \Phi_0(X,Y)$ with invertible end-points will be called an *admissible path.* In **[10]**, Jacobo Pejsachowicz and I introduced the following

Definition *The parity, $\sigma(L,I)$, of an admissible path L on I is defined by the formula*

$$\sigma(L,I) = \epsilon(M_a \circ L_a) \cdot \epsilon(M_b \circ L_b)$$

where $M: I \to GL(Y,X)$ is any parametrix for L.

Proposition 2.6 *The parity is properly defined.*
Proof. Let $L: I \to \Phi_0(X,Y)$ be a path with invertible end-points. By Theorem 2.5, this path has a parametrix. Suppose that $\bar{M}, M: I \to GL(Y,X)$ are two parametrices for $L: I \to \Phi_0(X,Y)$. Then $t \mapsto \bar{M}_t \circ (M_t)^{-1}, t \in I$ defines a path in $GL_C(X)$ joining $\bar{M}_a \circ (M_a)^{-1}$ to $\bar{M}_b \circ (M_b)^{-1}$. Proposition 2.2 implies that

$$\epsilon(\bar{M}_a \circ (M_a)^{-1}) = \epsilon(\bar{M}_b \circ (M_b)^{-1}).$$

Hence, since ϵ takes values in $\{\pm 1\}$, using Corollary 2.3 we conclude that

$$\epsilon(M_a \circ L_a) \cdot \epsilon(M_b \circ L_b) =$$

$$\epsilon(\bar{M}_a \circ (M_a)^{-1}) \cdot \epsilon(M_a \circ L_a) \cdot \epsilon(\bar{M}_b \circ (M_b)^{-1}) \cdot \cdot \epsilon(M_b \circ L_b) =$$

$$\epsilon(\bar{M}_a \circ L_a) \cdot \epsilon(\bar{M}_b \circ L_b).$$

Q.E.D.

Clearly the parity of a path of isomorphisms is 1 and the parity of an admissible path of compact vector fields is the product of the values of the function ϵ at the end-points of the path. We record the following properties for further reference **[10,15]**: the proofs of these are based on Proposition 2.2, Corollary 2.3 and Theorem 2.5.

The Homotopy Property: *The parity of an admissible path $L: I \to \Phi_0(X,Y)$ is 1 if and only if the path can be homotoped into $GL(X,Y)$ through a homotopy of admissible paths. In particular, the parity is a homotopy invariant of admissible paths.*

Multiplicativity Under Partitions of I: *If $L: I \to \Phi_0(X,Y)$ is admissible and if $L(c)$ is invertible at $c \in I$, then*

$$\sigma(L,I) = \sigma(L,[a,c]) \cdot \sigma(L,[c,b]).$$

Note that as a consequence of the homotopy property of the parity it follows that the parity is invariant under composition with a path of isomorphisms.

The parity also has significance as an intersection index. To describe this, I now turn to a description of a stratification of the singular Fredholm operators $\mathcal{S}(X,Y) = \Phi_0(X,Y) \setminus GL(X,Y)$ which closely parallels the corresponding description of $\mathcal{S}(R^n)$. Indeed, the set of singular linear Fredholm operators of index 0, $\mathcal{S}(X,Y) = \Phi_0(X,Y) \setminus GL(X,Y)$, can be decomposed as

$$\mathcal{S}(X,Y) = \bigcup_{k=1,\dots\infty} \mathcal{S}_k(X,Y),$$

where each $\mathcal{S}_k(X,Y)$ consists of those operators with a k-dimensional nullspace.

Theorem 2.7 ([14]) *The stratum $\mathcal{S}_1(X,Y)$ is a submanifold of $\Phi_0(X,Y)$ of codimension one. A C^1 path $L: R \to \Phi_0(X,Y)$ crosses $\mathcal{S}_1(X,Y)$ transversally at λ_0 if and only if it verifies the following property (T) at λ_0:*

$$\dim \text{Null } L_{\lambda_0} = 1$$

and (T)

$$L'_{\lambda_0}(\text{Null } (L_{\lambda_0})) \oplus \text{Im } (L_{\lambda_0}) = Y.$$

As in the finite dimensional case, under additional assumptions the transversal crossing of $\mathcal{S}_1(X,Y)$ can be more directly related to the behavior of the spectra of the path L. The proof of the first of the two following corollaries is clear.

Corollary 2.8 *Let $A \in L(X)$, and define $L: R \to L(X)$ by $L_\lambda = \lambda I - A$. Then $L: R \to L(X)$ crosses $\mathcal{S}_1(X)$ transversally at λ_0 if and only if λ_0 is a simple eigenvalue of A*

Corollary 2.9 *Let $L: R \to \Phi_0(X)$ be C^1 and suppose that $\lambda_0 \in R$ has the property that 0 is a simple eigenvalue of L_{λ_0}. Then $L: R \to \Phi_0(X)$ crosses $\mathcal{S}_1(X)$ transversally at λ_0 if and only if $\epsilon'(\lambda_0) \neq 0$, where I is a neighborhood of λ_0 and $\epsilon: I \to R$ is the unique C^1 path with $\epsilon(\lambda_0) = 0$ and $\epsilon(\mu)$ an eigenvalue of L_μ for each μ in I.*

Proof. Since 0 is a simple eigenvalue of L_{λ_0}, it is isolated in the complex spectrum of its complexification $L_{C,\lambda_0} \in \Phi_0(X_C)$. Choose Γ to be a circle in the complex plane, centered at the origin which isolates 0 from other points in the spectrum of L_{C,λ_0}. Then choose an interval I about λ_0 having the property that if $\lambda \in I$, then the spectrum of $L_{C,\lambda}$ does not intersect Γ. Define $P: I \to L(X_C)$ by setting P_λ equal to the spectral projection of $L_{C,\lambda}$ inside Γ. According to Corollary 2.1, the (complex) dimension of the image of these projections is constant, and, by simplicity of 0 as an eigenvalue of L_{λ_0}, this constant is 1. Fix x_0 to be any nonzero member of Null L_{λ_0} and define $z(\lambda) = P_\lambda(x_0 + i0)$ for $\lambda \in I$. Since $P_\lambda(X_C)$ has complex dimension 1 and $L_{C,\lambda}$ maps this space into itself, for each $\lambda \in I$ there are real numbers $\epsilon(\lambda)$ and $\mu(\lambda)$ such that $L_{C,\lambda}(z(\lambda)) = (\epsilon(\lambda) + i\mu(\lambda))z(\lambda)$. Clearly, $z(\lambda)$ is a C^1 function of λ and since $z(\lambda_0) \neq 0$ we may suppose that $z(\lambda) \neq 0$ for $\lambda \in I$. It follows that each $\mu(\lambda) = 0$, for otherwise $\epsilon(\lambda) \pm i\mu(\lambda)$ would be distinct eigenvalue of $L_{C,\lambda}$ which lie within Γ, and this contradicts the fact that the spectral multiplicity of $L_{C,\lambda}$ inside Γ is 1. Letting $x(\lambda)$ be the real part of $z(\lambda)$, we see that $x: I \to X$ is C^1, that $x(\lambda_0) \neq 0$ and that

$$L_\lambda(x(\lambda)) = \epsilon(\lambda)x(\lambda) \text{ for all } \lambda \in I.$$

Since 0 is a simple eigenvalue of L_{λ_0}, Null $L_{\lambda_0} \oplus$ Im $L_{\lambda_0} = X$. Choose ψ to be a continuous real linear functional on X which vanishes on Im L_{λ_0} and assumes the value 1 at x_0. By normalizing, we may assume that $\psi(x(\lambda)) \equiv 1$. Thus,

$$\psi(L_\lambda(x(\lambda))) = \epsilon(\lambda) \text{ for all } \lambda \in I.$$

Differentiating this identity, we obtain $\psi(L'_{\lambda_0}(x_0)) = \epsilon'(\lambda_0)$. Thus $L'_{\lambda_0}(x_0)$ is in the image of L_{λ_0} if and only if $\epsilon'(\lambda_0) = 0$. Since Null $L_{\lambda_0} \oplus$ Im $L_{\lambda_0} = X$ and the dimension of Null L_{λ_0} is 1, L satisfies property (T) at λ_0 if and only if $\epsilon'(\lambda_0) \neq 0$. **Q.E.D.**

Given a path $L: R \to \Phi_0(X,Y)$ which has λ_0 as an isolated singular point, the parity of the restriction of the path to a neighborhood of λ_0 which contains no other singular points will be denoted by $\sigma(L,\lambda_0)$.

Proposition 2.10 *Suppose that $L: R \to \Phi_0(X,Y)$ is a C^1 path which crosses $\mathcal{S}_1(X,Y)$ transversally at λ_0. Then $\sigma(L,\lambda_0) = -1$.*

Proof. Choose V to be a subspace of Y which is a linear complement in Y of Im L_{λ_0} and choose $Q \in L(Y)$ to be a projection onto V with $I - Q$ a projection onto Im L_{λ_0}. Then we may select an interval I about λ_0 and a C^1 path of projections $P: I \to L(X)$ such that Im P_λ = Null $(I - Q) \circ L_\lambda$ and $||P_\lambda - P_{\lambda_0}|| < 1$ for all $\lambda \in I$.* Since the parity is invariant under composition with paths of isomorphisms, by composing on the right with the path $\lambda \mapsto I - P_{\lambda_0} + P_\lambda$, choosing $S \in GL(\text{Null } L_{\lambda_0}, V)$ and then using $\lambda \mapsto [(I - Q) \circ L_\lambda + S \circ P_{\lambda_0}]^{-1}$ as a parametrix, it follows that $\sigma(L,\lambda_0) = \sigma(M,\lambda_0)$, where $M_\lambda = (I - P_{\lambda_0}) + S^{-1} \circ Q \circ L_\lambda$. Observe that M is a path of compact vector fields, so that $\sigma(M,\lambda_0)$ can be computed by counting negative real eigenvalues. But the only real eigenvalues of M_λ are 1 and $\eta(\lambda)$, where, fixing x_0 to be some nonzero member of Null L_{λ_0},

$$QL_\lambda(x_0) = \eta(\lambda)S(x_0), \quad \lambda \in I.$$

Thus $\sigma(L,\lambda_0) = -1$ if and only if $\eta(\lambda)$ changes sign as λ passes through λ_0. But it is clear that $\eta'(\lambda_0) \neq 0$ if and only if $QL'_{\lambda_0}(x_0) \neq 0$. According to Theorem 2.9, $L: R \to \Phi_0(X,Y)$ crosses $\mathcal{S}(X,Y)$ transversally at λ_0 if and only if $QL'_{\lambda_0}(x_0) \neq 0$. Thus $\sigma(L,\lambda_0) = -1$. **Q.E.D.**

By the multiplicative property of the parity with respect to partitions, if a C^1 admissible path of Fredholm operators has only a finite number of singular points, at each of which the path crosses $\mathcal{S}(X,Y)$ transversally, then from the preceding proposition we see that the parity is $(-1)^m$, where m is the number of singular points. The next result, which is the infinite dimensional correspondent of Theorem 1.12, asserts that this is the generic situation.

Theorem 2.11 ([14]) *If the set $\mathcal{S}(X,Y)$ of singular Fredholm operators of index 0 is decomposed as*

$$\mathcal{S}(X,Y) = \bigcup_{k=1,\ldots\infty} \mathcal{S}_k(X,Y),$$

where each $\mathcal{S}_k(X,Y)$ consists of those operators with a k-dimensional nullspace, then each $\mathcal{S}_k((X,Y)$ is a submanifold of $\Phi_0(X,Y)$ of codimension k^2. Moreover, given an

---

* Indeed, observe that after choosing P_{λ_0} to be a projection onto Null L_{λ_0}, selecting an operator $S \in GL(\text{Null } L_{\lambda_0}, V)$ and defining $T_\lambda = (I-Q) \circ L_\lambda + S \circ P_{\lambda_0}$, the formula $P_\lambda = T_\lambda^{-1} \circ T_{\lambda_0} \circ P_{\lambda_0} \circ T_{\lambda_0}^{-1} \circ T_\lambda$ defines such a family of projections, on a sufficiently small interval about λ_0.

admissible path $L: I \to \Phi_0(X, Y)$, there are arbitrarily close smooth perturbations of L which cross $\mathcal{S}(X, Y)$ transversally and

$$\sigma(L, I) = (-1)^m, \tag{2.12}$$

where m is the number of singular points of any such close perturbation.

The parity of a path of linear compact vector fields with invertible ends is determined by its values of the end points. In general, this is not the case. In fact, there are closed paths of linear Fredholm operators whose parity is -1. In view of the preceding theorem, this means that there are closed paths of linear Fredholm operators which intersect the singular set transversally an *odd* number of times. This leads to quite striking bifurcation behavior for families of nonlinear mappings parametrized by parameter spaces with nontrivial topology (see §5). To explain how such paths arise, we recall the following seminal result.

Kuiper's Theorem ([24]) *Let X be a separable infinite dimensional real Hilbert space. Then $GL(X)$ is contractible.*

It is now also known that real nonseparable Hilbert spaces, that the real Sobolev spaces and the real Hölder spaces also have contractible general linear groups (cf. **[27]**). It is convenient to call a real Banach space having the property that $GL(X)$ is contractible a Kuiper space.

If X is a Kuiper space, then, of course, $GL(X)$ is connected. Moreover, $\Phi_0(X)$ inherits connectedness from $GL(X)$. To see this, just observe that if $T \in \Phi_0(X)$ and S is a parametrix for T, then $t \mapsto tT + (1-t)S^{-1}, 0 \le t \le 1$ defines a path in $\Phi_0(X)$ connecting T to $GL(X, Y)$. Also, if X is a Kuiper space, then, of course, $GL(X)$ is simply connected. *But $\Phi_0(X)$ does not inherit simple connectedness from $GL(X)$.* This is not immediately clear. But it is clear why the preceding argument will not work. Namely, given a family $L: S^1 \to \Phi_0(X)$, since S^1 is not contractible we cannot invoke Theorem 2.5 in order to conclude that it has a parametrix.

In fact, if X is a Kuiper space, then there are families $L: S^1 \to \Phi_0(X)$ which do not have parametrices and these are precisely the closed paths in $\Phi_0(X)$ which cannot be deformed in $\Phi_0(X)$ to a constant path. The parity provides a simple way of verifying this assertion. Actually, to do this it is first useful to define the parity for general closed paths.

Definition *The parity $\sigma(L, S^1)$ of a closed path $L: S^1 \to \Phi_0(X)$ is defined by the formula*

$$\sigma(L, S^1) = \epsilon(M_0^{-1} \circ M_1),$$

where $\psi: [0,1] \to S^1$ is any simple parametrization of the circle and $M: [0,1] \to GL(X)$ is any parametrix for the path $t \mapsto L_{\psi(t)}, 0 \le t \le 1$.

Using Proposition 2.2 and Theorem 2.5, one can verify the following

Theorem 2.11 ([15]) *The parity of a closed path of linear Fredholm operators is properly defined. Moreover, for a closed path $L: S^1 \to \Phi_0(X)$, the following assertions are equivalent:*

(i) The parity of $L: S^1 \to \Phi_0(X)$ is 1.

(ii) There is a parametrix for the family $L: S^1 \to \Phi_0(X)$.
(iii) $L: S^1 \to \Phi_0(X)$ *can be deformed in* $\Phi_0(X)$ *into GL(X).*

Of course, the above equivalences do not guarantee that there are any closed paths of Fredholm operators having parity -1. Certainly, if $X = R^n$, then $\Phi_0(R^n) = L(R^n)$ is contractible, and so there are no such paths. However, if X is a Kuiper space, then there are such paths. This is not difficult to see. Indeed choose L_0 and L_1 to be any two operators in $GL_C(X)$ with $\epsilon(L_0) = -\epsilon(L_1)$. Since $GL(X)$ is connected, L_0 and L_1 may be joined by a path $L: [0,1] \to GL(X)$. On the other hand, the path $(2-t)L_1 + (t-1)L_0, 1 \leq t \leq 2$ defines a path of linear compact vector fields joining L_1 to L_0. The closed path $L: [0,2] \to \Phi_0(X)$ defines a path $L: S^1 \to \Phi_0(X)$, which, by the multiplicative property of the parity under partitions, has parity -1. Another way of obtaining closed paths with parity -1 is to consider a path η with invertible ends which intersects the singular set once and transversally. Then just close the path η by joining the ends by a path in $GL(X,Y)$ to obtain a closed path of parity -1.

Observe that Theorem 2.11 implies that the parity of a closed path is a *free* homotopy invariant of the path, and so the example just described shows that the Poincaré Group of $\Phi_0(X)$ is nontrivial if X is a Kuiper space. In fact, it is not difficult to see that the properties of the parity of a closed path may be synthasised as the following assertion concerning the Poincaré Group of $\Phi_0(X)$. Define $\Pi_1(\Phi_0(X))$ to be the Poincaré group of $\Phi_0(X)$ with the identity operator chosen as base point.

Theorem 2.12 ([13]) *Suppose that X is a Kuiper space. For a homotopy class* $[\alpha]$ *in* $\Pi_1(\Phi_0(X))$, *define* $\sigma([\alpha]) = \sigma(\alpha, S^1)$. *Then*

$$\sigma: \Pi_1(\Phi_0(X)) \to Z_2$$

is a group isomorphism.

§3 Local Bifurcation for One Parameter Families

We now turn to bifurcation for families of mappings between the real Banach spaces X and Y. Given a parameter space Λ and a family $f: \Lambda \times X \to Y$ of nonlinear Fredholm mappings parametrized by Λ, we will consider bifurcation for the solutions of the equation

$$f(\lambda, x) = 0, \qquad (\lambda, x) \in \lambda \times X. \tag{3.1}$$

Assume that $f(\lambda, 0) = 0$ for all $\lambda \in R$, and call $\Lambda \times \{0\}$ the *branch of trivial solutions* of equation (3.1). A point $(\lambda, 0)$ in the trivial branch is called a *local bifurcation point* for equation (3.1) provided that every neighborhood of $(\lambda, 0)$ contains nontrivial solutions of the equation. We also assume that $D_x f(\lambda, 0) \equiv L_\lambda$ exists as a Fréchet derivative at each point in the trivial branch. Denote by Bif (f) the set of local bifurcation points of equation (3.1), and let $\Sigma = \{\lambda | L_\lambda \text{ is singular }\}$ be the set of singular points of the family of linearizations. There is the following classical necessary condition on the family of linearizations in order for bifurcation to occur.

Lemma 3.2 *Assume that* $f(\lambda, x) = L_\lambda(x) + R(\lambda, x)$, *where R vanishes to first order in x at x=0, uniformly on bounded sets of* λ. *Assume also that* $L: \Lambda \to \Phi_0(X, Y)$ *is continuous. Then*

$$Bif\,(f) \subset \Sigma.$$

Proof. If $\{(\lambda_k, x_k)\}$ is a sequence of nontrivial solutions of equation (3.1) which converges to $(\lambda_0, 0)$, then $\{L_{\lambda_0}(v_k)\}$ converges to 0, where each v_k is the normalization of x_k. Since L_{λ_0} is Fredholm of index 0, its restriction to closed bounded sets is proper, and so a subsequence of $\{v_k\}$ converges to a unit vector v having the property that $L_{\lambda_0}(v) = 0$. Thus λ_0 is in Σ. **Q.E.D.**

Because of this lemma, the points of Σ are usually referred to as *potential bifurcation points* of equation (3.1). Of course, potential bifurcation points need not, in fact, be bifurcation points *.

In this section will only consider the case $\Lambda = R$. In a neighborhood of an isolated potential bifurcation point, we will describe a necessary and sufficient condition on the family of linearizations in order that it actually be a bifurcation point, and also describe necessary and sufficient conditions on the family of linearizations in order that the set of bifurcating nontrivial solutions comprise a curve. We begin with the latter.

It turns out that the following very simplest situation is indicative of the most general case. Assume that $X = Y = R$ and $f: R^2 \to R$ is analytic. In view of Lemma 3.2, in order for λ_0 to be a bifurcation point of equation (3.1) it is necessary that $\partial f/\partial x(\lambda_0, 0) = 0$. Because of the presence of the trivial solutions, f factors as $f(\lambda, x) = xg(\lambda, x)$ for all $(\lambda, x) \in R^2$, where g is also analytic. Assume that $\partial f/\partial x(\lambda_0, 0) = 0$, which just means that $g(\lambda_0, 0) = 0$. If $\partial g/\partial x(\lambda_0, 0) \neq 0$, then the Implicit Function Theorem implies that near to $(\lambda_0, 0)$ the solutions of $g(\lambda, x) = 0$ comprise the graph of a smooth path $\lambda = \eta(x), x \in I$ with $\eta(0) = \lambda_0$, and so in a neighborhood of $(\lambda_0, 0)$ the nontrivial solutions of equation (3.1) are the points (λ, x) of the form $\lambda = \eta(x), x \in I \setminus \{0\}$. Of course, the assumption that $\partial g/\partial \lambda(\lambda_0, 0) \neq 0$ just means that $\partial^2 f/\partial\lambda\partial x(\lambda_0, 0) \neq 0$.

The above argument of "dividing out the zeroes" does not require analyticity and works for maps $f: R \times X \to Y$.

The Simple Eigenvalue Theorem ([3]) *Suppose that* $f: R \times X \to Y$ *is* C^2 *and that* $f(\lambda, 0) \equiv 0$. *Assume that at* λ_0 *the family of linearizations satisfies the following property:*

$$\dim \text{Null}\, L_{\lambda_0} = 1$$

and (*T*)

$$L'_{\lambda_0}(\text{Null}\,(L_{\lambda_0})) \oplus \text{Im}\,(L_{\lambda_0}) = Y.$$

Then there is a neighborhood I of 0 in R and C^1 *functions* $\eta: I \to R$ *and* $\varphi I \to X$ *with* $\eta(0) = 0, \varphi(0) = 0$ *and* $\varphi'(0) \in$ *Null* $L_{\lambda_0} \setminus \{0\}$, *which have the property that in a*

---

* For instance, define $f: R^3 \to R^2$ by $f(\lambda, x, y) = (\lambda x - y(x^2 + y^2), \lambda y + x(x^2 + y^2))$. One immediately checks that $(x, y) = (0,0)$ is a solution of equation (3.1) for all λ, that $L_\lambda = \lambda I$, for each λ, so that the linearization with respect to (x, y) at (0,0) is singular at $\lambda_0 = 0$, but that this equation has no bifurcation points.

neighborhood of $(\lambda_0, 0)$ *the nontrivial solutions of equation (3.1) are the points* (λ, x) *of the form* $(\eta(s), \varphi(s)), s \in I \setminus \{0\}$.

Proof. Let x_0 be any nonzero member of Null L_{λ_0}, and let X_1 be a closed subspace of X which is a linear complement of Null L_{λ_0} in X. Define $g: R \times R \times X_1 \to Y$ by

$$g(\lambda, s, v) = \begin{cases} s^{-1} \cdot f(\lambda, s(x_0 + v)) & \text{if } s \neq 0 \\ D_x f(\lambda, 0)(x_0 + v) & \text{if } s = 0. \end{cases}$$

It is easy to check that $g: R \times R \times X_1 \to Y$ is C^1 and that $g(\lambda_0, 0, 0) \equiv 0$. Furthermore, since Null L_{λ_0} is spanned by x_0, the second part of assumption (T) is equivalent to the assertion that

$$D_{(\lambda,v)} g(\lambda_0, 0, 0) \in L(R \times X_1, Y) \text{ is an isomorphism.}$$

From the Implicit Function Theorem it follows that there is a neighborhood I of 0 in R and C^1 functions $\eta: I \to R$ and $\psi: I \to X_1$ having the property that there is a neighborhood of $(\lambda_0, 0)$ in $R \times X$ in which the only zeroes of g are those of the form $(\eta(s), s[x_0 + \psi(s)]), s \in I$. Defining $\varphi(s) = s\psi(s)$, the conclusion now follows from the form of g. **Q.E.D.**

Corollary 3.3 *Let* $X = Y$ *and suppose that* $f: R \times X \to X$ *is* C^2 *and that* $f(\lambda, 0) \equiv 0$. *Furthermore, suppose that the point* λ_0 *in R has the property that 0 is a simple eigenvalue of* L_{λ_0}. *Assume, moreover, that* $\epsilon'(\lambda_0) \neq 0$, *where I is a neighborhood of* λ_0 *and* $\epsilon: I \to R$ *is the unique* C^1 *path with* $\epsilon(\lambda_0) = 0$ *and* $\epsilon(\mu)$ *an eigenvalue of* L_μ *for each* μ *in I. Then there is a neighborhood I of 0 in R and* C^1 *functions* $\eta: I \to R$ *and* $\varphi I \to X$ *with* $\eta(0) = 0, \varphi(0) = 0$ *and* $\varphi'(0) \in$ *Null* $L_{\lambda_0} \setminus \{0\}$, *which have the property that in a neighborhood of* $(\lambda_0, 0)$ *the nontrivial solutions of equation (3.1) are the points* (λ, x) *of the form* $(\eta(s), \varphi(s)), s \in I \setminus \{0\}$.

Proof. Corollary 2.11 asserts that the assumptions on the path $\lambda \mapsto L_\lambda$ near λ_0 imply that the path satisfies property (T). So we may invoke the Simple Eigenvalue Theorem. **Q.E.D.**

The first explicit assertion of the Simple Eigenvalue Theorem which I know of, even in the finite dimensional case, is due to Crandall and Rabinowitz **[3]** (cf. also **[21]** and **[28]**). However, fifty years ago, in his paper **[18]** on the bifurcation of periodic orbits from a family of equilibria of a dynamical system, E. Hopf asserted the validity of Corollary 3.3 in the case $X = Y = R^n$ and f is analytic.

Transversal Crossing and Finite Algebraic Multiplicity: As we pointed out in the previous section, if $X = Y$ and if $L_\lambda = I - \lambda A$, for some fixed $A \in L(X)$, then L has property (T) at λ_0 if and only if λ_0 is a simple eigenvalue of A. Presumably because of this, the name "simple-eigenvalue bifurcation" has been attached to the above bifurcation theorem. However, it is useful to note that, even in the finite dimensional case, in the absence of extra assumptions, assumption (T) has no bearing on the algebraic multiplicity of 0 as an eigenvalue of L_{λ_0}. In general, λ_0 is an eigenvalue of L_{λ_0} of finite algebraic multiplicity if and only if L_{λ_0} is Fredholm of index 0 and λ_0 is an isolated singular point of the path $\lambda \mapsto (\lambda - \lambda_0) - L_{\lambda_0}, \lambda \in R$ (cf. **[14]**). On the other hand, property (T) can hold at λ_0 and yet 0 can be an eigenvalue of L_{λ_0} of infinite algebraic multiplicity*.

Transversal Crossing and Crossing of Negative Eigenvalues: In the finite dimensional case, or in the infinite dimensional case for paths of compact vector fields, it follows from Proposition 2.12 that if λ_0 is an isolated singular point of the path $L: R \to \Phi_0(X)$ which crosses $\mathcal{S}_1(X)$ transversally at λ_0, then

$$(-1)^{n_-} \cdot (-1)^{n_+} = -1,$$

where $n_\pm$ is the number of negative eigenvalues of $L_{\lambda_0 \pm \eta}$ and η is sufficiently small. So, in this case, property (T) certainly means that there has been a "crossing of negative eigenvalues" of L_λ as λ passes through λ_0. Moreover, Corollary 2.11 is an explicit assertion of the crossing of one real eigenvalue if 0 is a simple eigenvalue of L_{λ_0}. In the general infinite dimensional case, a C^1 path can cross $\mathcal{S}(X,Y)$ transversally and yet there is no sense in which there has been any crossing of negative eigenvalues. In fact, the path described in the footnote below has the property that if $\lambda \neq 0$, then L_λ has no negative eigenvalues and yet the path crosses the singular set transversally at 0. Both the parity of a path and the transversal crossing of $\mathcal{S}_1(X,Y)$ by the path are invariant under pointwise composition of the path with a path of isomorphisms, whereas, of course, spectral data can be destroyed by such compositions.

The Necessity of Property (T): Suppose that $L: R \to \Phi_0(X,Y)$ is C^2. Then in order that for every C^2 mapping $f: R \times X \to Y$ having the property that $f(\lambda, 0) \equiv 0$ and $D_x f(\lambda, 0) \equiv L_\lambda$ the set of nontrivial solutions of equation (3.1) in a neighborhood of $(\lambda_0, 0)$ comprise a smooth curve passing through $(\lambda_0, 0)$ it is necessary and sufficient that the path $L: R \to \Phi_0(X,Y)$ have property (T) at λ_0. Of course, the sufficiency is just the assertion of the Simple Eigenvalue Theorem. To verify the necessity, the important case is when $X = Y = R$. We argue by contradiction. Suppose that $\ell: R \to R$ is C^2 and that $\ell(\lambda_0) = \ell'(\lambda_0) = 0$. Write $\ell(\lambda) = \lambda^2 \phi(\lambda)$, where $\phi: R \to R$ is C^2. Then observe that

$$f(\lambda, x) = \ell(\lambda)x - x^3 \phi(\lambda) \text{ for all } (\lambda, x) \in R^2$$

defines a C^2 map with $f(\lambda, 0) \equiv 0$ and $D_x f(\lambda, 0) \equiv \ell_\lambda$ and yet there are two smooth curves of nontrivial solutions passing through λ_0. In the general case, if $L: R \to \Phi_0(X,Y)$ does not satisfy property (T) at λ_0, then if Null L_{λ_0} is more than one dimensional there are infinitely many bifurcating nontrivial branches of solutions of the equation $L_\lambda x = 0$. So one can suppose that Null L_{λ_0} is one dimensional, and also, by composing with a linear isomorphism on the right, that L_{λ_0} is a projection P having nullspace spanned by the vector x_0. Choose an interval I about λ_0 in which the restriction of $P \circ L_\lambda$ to the image of P has an inverse S_λ. Define $\ell(\lambda)t$ to be the x_0 coefficient of $t[(I-P) \circ L_\lambda](x_0 - S_\lambda P L_\lambda x_0)]$. Observe that if $\lambda \in I$, $t \in R$ and $\bar{x} \in P(X)$, then

---

* Let X be a separable Hilbert space with orthonormal basis $\{e_k\}_{k=-\infty}^{k=+\infty}$. Define $A, B \in L(X)$ by $A(e_i) = e_{i-1}, B(e_i) = 0$ if $i \neq 1, A(e_1) = 0, B(e_1) = e_0$. Set $L_\lambda = A + \lambda B$ for $\lambda \in R$. At $\lambda_0 = 0$, property (T) holds and yet 0 does not have finite algebraic multiplicity as an eigenvalue of A. Furthermore, if $\lambda \neq 0, L_\lambda$ has no negative eigenvalues.

$L_\lambda(tx_0 + \bar{x}) = 0$ if and only if $\ell(\lambda)t = 0$. But note that $\ell(\lambda_0) = \ell'(\lambda_0) = 0$ and so by the scalar case already considered, we may select a C^2 scalar map $\epsilon: I \times R \to R$ such that if $f(\lambda, t) = \ell(\lambda, t) + \epsilon(\lambda, t)$, then $f(\lambda, 0) \equiv 0$ and $D_x f(\lambda, 0) \equiv \ell(\lambda)$ and yet there are two smooth curves of nontrivial solutions of $f(\lambda, t) = 0$ passing through λ_0. Then it is clear that $F(\lambda, tx_0 + \bar{x}) = L_\lambda(tx_0 + \bar{x}) + \epsilon(\lambda, t)x_0$ defines a C^2 map having the property that $F(\lambda, 0) \equiv 0$ and $D_x F(\lambda, 0) = L_\lambda$ and there are two smooth curves of nontrivial solutions of $F(\lambda, t) = 0$ which pass through $(\lambda_0, 0)$. In [**21**], Kopell and Howard first showed, in the finite dimensional case, that property (T) was necessary in order to have the conclusions of the Simple Bifurcation Theorem for all higher order perturbations.

So, at an isolated singular point λ_0 of the path $L: R \to \Phi_0(X, Y)$, property (T), or equivalently, the transversal crossing of the set of singular operators, is the only data on the linearizations which will guarantee that the nontrivial solutions of equation (3.1) near $(\lambda_0, 0)$ comprise a smooth curve. However, transversal crossing is not a necessary condition for bifurcation. In fact, in the finite dimensional case, to ensure bifurcation at a point λ_0 where $\det L_\lambda = 0$ it is not necessary to require that $(d/d\lambda)(\det L_\lambda)|_{\lambda=\lambda_0} \neq 0$: rather, it suffices just that $\det L_\lambda$ change sign as λ passes through λ_0. This is the content of the following Local Bifurcation Theorem of Krasnosel'skii :

Theorem 3.4 ([22]) *Suppose that λ_0 is an isolated singular point of the path of linearizations $L: R \to L(R^n)$. Assume that*

$$\text{sgndet } L_\lambda \text{ changes sign at } \lambda_0. \tag{C}$$

Then λ_0 is a bifurcation point for equation (3.1).

Proof. If the conclusion is false, then there is a product neighborhood $I \times U$ of $(\lambda_0, 0)$ in which the only solutions of equation (3.1) are the trivial ones. Fix a and b in I, with $a < \lambda_0 < b$. Then if U is replaced by a small enough ball about the origin, it follows from the homotopy invariance of the Brouwer degree for maps on this ball that sgn det L_a = sgn det L_b **Q.E.D.**.

Theorem 3.4 is also sharp. In [**20**], Ize showed that if $L: R \to L(R^n)$ is C^1, λ_0 is an isolated singular point of this path and if sgn det (L_λ) does not change sign at λ_0, then there is a C^1 function $f: R \times R^n \to R^n$ for which $f(\lambda, 0) \equiv 0$, $D_x f(\lambda, 0) \equiv L_\lambda$ and for which λ_0 is not a bifurcation point of equation (3.1). In the analytic case, a similar result was established by Esquinas [**6**] in the framework of the notion of generalized multiplicity.

In the infinite dimensional case, there is a large literature on sufficient conditions on the family of linearizations L_λ, at an isolated singular point $\lambda_0 \in \Sigma$, other than the transversal crossing of $\mathcal{S}(X, Y)$, which will imply local bifurcation of the nontrivial solutions of (3.1) from $(\lambda_0, 0)$ (cf. [**22**], [**26**], [**34**], [**19**], [**25**], [**2**], [**23**], [**33**], [**20**], [**7**], [**5**], [**6**], [**30**] and the references therein). In [**12**], a certain unity was imposed on this seemingly disparate body of local bifurcation results by proving that the parity is a complete linear bifurcation invariant for one parameter problems. More precisely, using the invariance of the parity under Lyapunov-Schmidt Reduction and the finite dimensional results mentioned above, at an isolated singular point of the path of linearizations one has

Theorem 3.5 *Let $\Lambda = R$ and $L: R \to \Phi_0(X,Y)$ be a C^1 family of linear Fredholm operators for which $\lambda_0 \in R$ is an isolated singular point. Then $\sigma(L, \lambda_0) = -1$ if and only for every C^1 family of nonlinear Fredholm mappings $f: R \times X \to Y$ with $Df_x(\lambda, 0) = L_\lambda$ and $f(\lambda, x) = 0$ for all $\lambda \in R$ there is local bifurcation from λ_0 of nontrivial solutions of the equation $f(\lambda, x) = 0$.*

According to Proposition 2.12, at points where a curve crosses $\mathcal{S}(X,Y)$ transversally, the parity is -1. In this case, however, rather than the above theorem, one can apply the much stronger Simple Eigenvalue Theorem. In the context of local bifurcation, the above theorem is of interest when the path of linearizations crosses $S(X,Y)$ degenerately. The many sufficient conditions on the path of linearizations for bifurcation at isolated potential bifurcation points can, from the perspective provided by Theorem 3.5, be seen as being specific calculations of the parity for paths which are crossing $S(X,Y)$ degenerately. Many of these are phrased as generalizations, to the case of the linearizations of general one parameter families, at isolated singular points, of the concept of algebraic multiplicity for an eigenvalue of a matrix (cf. [4], [**26**], [**19**], [**5**], [**6**] and [**30**] for some of these generalizations). This was considered in detail in [**14**].

It is of interest to compute $\sigma(L, \lambda_0)$ at degenerate isolated singular points. In [**14**], it is shown that the following assumptions (i) and (ii) each imply that $\sigma(L, \lambda_0) = -1$, and so we have the following corollary of Theorem 3.4.

Corollary 3.6 *Suppose that $f: R \times X \to Y$ is C^1 and that $f(\lambda, x) = 0$ for all $\lambda \in R$. Define $Df_x(\lambda, 0) = L_\lambda$ for all $\lambda \in R$ and suppose that $L: R \to \Phi_0(X,Y)$ is differentiable at λ_0. Then λ_0 is a local bifurcation point of equation (3.1) provided that one of the following assumptions hold:*

(i)

$$L'_{\lambda_0}(\textit{ Null } L_{\lambda_0}) \oplus \textit{ Im } L_{\lambda_0} = Y$$

and

$$\dim \textit{Null } L_{\lambda_0} \textit{ is odd.}$$

(ii)

$$X = Y, \quad L_\lambda L_{\lambda_0} = L_{\lambda_0} L_\lambda \textit{ for all } \lambda \in R, \quad \textit{Null } L_{\lambda_0} \cap \textit{ Null } L'_{\lambda_0} = \{0\}$$

and

$$0 \textit{ is an eigenvalue of } L_{\lambda_0} \textit{ of odd algebraic multiplicity.}$$

In the case that $X = Y = R^n$, the first part of assumption (i) means that the first integer j for which $(d/d\lambda)^j(\det L_\lambda)|_{\lambda=\lambda_0} \neq 0$ is k, where $k =$ dim Null L_{λ_0}. So the sign of the determinate changes if and only if k is odd. In [**34**], Westreich introduced assumption (i) as a sufficient condition for local bifurcation for a path of nonlinear Fredholm maps.

In the case that L_λ has the particular form $L_\lambda = \lambda I - A$, for some fixed $A \in L(X)$, the first part of assumption (ii) is satisfied and the second part means that λ_0 is an eigenvalue of A of odd algebraic multiplicity. In fact, this is exactly the form of the data on the path of linearizations of a family of compact vector fields which was presented in the Rabinowitz Bifurcation Theorem [**31**]. For general paths of compact vector fields, in [**33**] assumption (ii) was introduced by Toland as a sufficient condition to guarantee a change of sign in the Leray-Schauder degree and thus ensure bifurcation.

§4 Degree for Quasilinear Fredholm Mappings

In [**32**], Snirel'man introduced the class of quasilinear Fredholm maps in order to study the nonlinear Riemann-Hilbert problem. He defined a topological degree for such maps but, because of various choices which are made in the construction, his degree is only defined up to sign. While this is perfectly adequate for the existence problems studied in [**32**], it is not suitable for the study of multiplicity and bifurcation problems.

In [**10,15**], Jacobo Pejsachowicz and I developed an additive, integer-valued degree theory for the class of quasilinear Fredholm mappings. Since this class includes all of $GL(X,Y)$, in view of our description in §2 of the topology of $GL(X,Y)$ and $\Phi_0(X,Y)$, it is clear that such a degree cannot possess the property of homotopy invariance in the usual sense. It was in order to describe the homotopy property of this degree that we first introduced the notion of parity. In fact, the parity provides a complete description of the possible changes in sign of the degree and thereby enabled us to use the degree to prove multiplicity and bifurcation theorems for quasilinear Fredholm mappings. Since within this framework one can study general fully nonlinear elliptic boundary value problems (cf. [**1**], [**10,15**]), we were able to describe global bifurcation for quite general families of fully nonlinear elliptic boundary value problems. Moreover, because of the extra richness in the topology of $\Phi_0(X,Y)$, for instance the fact that it may not be simply connected, we were able to use this degree and the parity to describe new bifurcation phenomena for families of nonlinear maps parametrized by S^1. I will devote this section to sketching the construction and some of the properties of the degree developed in [**10,15**]. Then the final section will be devoted to using this degree to establish global bifurcation results.

We assume throughout this section that X and Y are real Banach spaces, and that $\bar{X}$ is another real Banach space in which X is compactly embedded.

Definition *A mapping $f: X \to Y$ it is called quasilinear Fredholm provided that f has a representation of the form*

$$f(x) = L(x)x + C(x) \text{ for } x \in X, \tag{4.1}$$

where

(i) $L: X \to \Phi_0(X,Y)$ is the restriction to X of a continuous map

$$\bar{L}: \bar{X} \to \Phi_0(X,Y),$$

and

(ii) $C: X \to Y$ is compact.

We will refer to formula (4.1), where (i) and (ii) are satisfied, as a *representation* of f, and call the family $L: X \to \Phi_0(X,Y)$ a *principal part* of f.

Since $\bar{X}$ is a contractible metric space, according to Theorem 2.5, every family of linear Fredholm operators parametrized by $\bar{X}$ has a parametrix. Using such a parametrix for the above $\bar{L}: \bar{X} \to \Phi_0(X, Y)$ and the compact embedding of $\bar{X}$ in X, it is not difficult to verify

Proposition 4.2 *Let $f: X \to Y$ be quasilinear Fredholm. Then f may be represented as*

$$f(x) = M(x)(x - \psi(x)) \text{ for } x \in X, \tag{4.3}$$

where $\bar{M}: \bar{X} \to GL(X, Y)$ is a family of isomorphisms and $\psi: X \to X$ is compact.

Of course, if f is represented by (4.3), then the zeroes of f coincide with the zeroes of the compact vector field $I - \psi$. Thus, if $\mathcal{O}$ is an open, bounded subset of X on the boundary of which f does not vanish, then $I - \psi$ is also nonvanishing on the boundary of $\mathcal{O}$ and hence its Leray-Schauder degree, $\deg_{L.S.}(I - \psi, \mathcal{O}, 0)$, is defined. One cannot define the degree of f on $\mathcal{O}$ to be $\deg_{L.S.}(I - \psi, \mathcal{O}, 0)$, because it turns out that the sign of $\deg_{L.S.}(I - \psi, \mathcal{O}, 0)$ depends on the choice of representation (4.3).

In order to determine a coherent choice of sign, it is useful to first consider the case when f is a linear isomorphism and $\mathcal{O}$ is the unit ball. As motivation for how one might proceed, recall Proposition 1.1 and Proposition 2.2 regarding the the roles of the Brouwer and Leray-Schauder degrees, respectively, in labeling connected components of certain sets of linear isomorphisms.

If X and Y are of the same finite dimension, a choice of orientation of X and of Y defines the determinant, $\det(T)$, for $T \in GL(X, Y)$. According to Proposition 1.1, $\epsilon: GL(X, Y) \to \{\pm 1\}$, defined by $\epsilon(T) = \text{sgn} \det(T)$, labels the two connected components of $GL(X, Y)$. Of course, $\epsilon(T)$ is the Brouwer degree of T with respect to the choice of orientations.

If X is infinite dimensional, then according to Proposition 2.2, $GL_C(X)$, the group of compact vector fields in $GL(X)$, also has two components, which are labeled by the function $\epsilon: GL_C(X) \to \{-1, +1\}$ defined by setting $\epsilon(T) = (-1)^m$, where m is the sum of the algebraic multiplicities of the negative eigenvalues of T. In this case, $\epsilon(T)$ is the Leray-Schauder degree of T.

For general spaces X and Y, while $GL(X, Y)$ may be connected, we may divide $GL(X, Y)$ into equivalence classes under the Calkin equivalence relation, $T \approx S$ if $T - S$ is compact, and then observe that given $T \in GL(X, Y)$, the correspondence $S \mapsto T \circ S$ is a continuous bijection between $GL_C(X)$ and the equivalence class, with respect to the Calkin equivalence relation, to which T belongs. Hence, since $GL_C(X)$ has two connected components, so does each equivalence class. The Leray-Schauder degree distinguishes the components of the equivalence class of the identity in $GL(X)$. It is reasonable to require of any degree for quasilinear Fredholm maps that it distinguish the components of each Calkin equivalence class. Observe that, in view of the preceding correspondence, Proposition 2.2 and Corollary 2.3, if T and S in $GL(X, Y)$ are equivalent, then they lie in the same component of their equivalence class if and only if $\deg_{L.S.}(T^{-1}S) = 1$. Accordingly, we make the following

Definition *Let X and Y be real Banach spaces. A map $\epsilon: GL(X, Y) \to \{\pm 1\}$ is called an orientation provided that whenever $T, S \in GL(X, Y)$ are equivalent, then*

$$\epsilon(T) \cdot \epsilon(S) = \deg_{L.S.}(T^{-1}S).$$

An orientation distinguishes the components of each Calkin equivalence class. We always insist that an orientation of $GL(X)$ assign 1 to the identity. This normalization assumption means that an orientation of $GL(X)$ assigns the Leray-Schauder degree to each member of $GL_C(X)$.

Once an orientation ϵ of $GL(X,Y)$ is chosen, the degree is defined as follows.

Definition *Suppose that $f: X \to Y$ is a quasilinear Fredholm map and $\mathcal{O}$ is an open, bounded subset of X on the boundary of which f does not vanish. Let ϵ be an orientation of $GL(X,Y)$. Then the degree of f on $\mathcal{O}$, with respect to the orientation ϵ, is defined by*

$$\deg_\epsilon(f, \mathcal{O}, 0) = \epsilon(M(0)) \deg_{L.S.}(I - \psi, \mathcal{O}, 0), \tag{4.4}$$

where M and ψ are as in (4.3).

From the observation that the principal parts of two different representations of a quasilinear Fredholm map differ by a family of compact operators and the composition property of the Leray-Schauder degree, it follows that the right-hand side of (4.4) is independent of the representation (4.3), and so the degree is properly defined.

The degree has the usual additivity, existence and Borsuk-Ulam properties. In the case that $X = Y$ and f is a compact vector field, it agrees with the Leray-Schauder degree. The most interesting aspects of this degree are the homotopy formula and the regular-value formula. I will only describe two particular cases of these.

Definition *If Λ is any topological space, then a mapping $F: \Lambda \times X \to Y$ is called a family of quasilinear Fredholm maps parametrized by Λ provided that*

$$F(\lambda, x) = L_\lambda(x)x + C(\lambda, x) \text{ for } \lambda \in \Lambda \text{ and } x \in X,$$

where

$$\bar{L}: \Lambda \times \bar{X} \to \Phi_0(X,Y) \text{ is continuous and } C: \Lambda \times X \to Y \text{ is compact.}$$

If $\Lambda = [0,1]$ such a family is called a quasilinear Fredholm homotopy.

A Homotopy Formula *Let $H: [0,1] \times X \to Y$ be a quasilinear Fredholm homotopy which also has the property that each $H(t,\cdot) \equiv H_t$ is Fréchet differentiable at $x = 0$ and $t \mapsto L_t \equiv D_x H_t(0), t \in [0,1] \equiv I$, defines a continuous path with invertible end-points. Let $\mathcal{O}$ be an open, bounded subset of $I \times X$ such that each H_t does not vanish on the boundary of $\mathcal{O}_t$. Let ϵ be an orientation of $GL(X,Y)$. Then*

$$\deg_\epsilon(H_0, \mathcal{O}_0, 0) = \delta \deg_\epsilon(H_1, \mathcal{O}_1, 0), \tag{4.5}$$

where

$$\delta = \deg_\epsilon(L_0) \cdot \sigma(L, I) \cdot \deg_\epsilon(L_1). \tag{4.6}$$

The proof of the above formula is not difficult. Indeed, one can choose a representation of H of the form

$$M_t(x)(x - \psi(t,x)) \text{ for } t \in I \text{ and } x \in X,$$

where

$$\bar{M}: [0,1] \times \bar{X} \to GL(X,Y) \text{ is continuous and } \psi: [0,1] \times X \to Y \text{ is compact.}$$

From the homotopy invariance of the Leray-Schauder degree and the definition of the quasilinear Fredholm degree, it follows that

$$\deg_\epsilon(H_0, \mathcal{O}_0, 0) = \epsilon(M_0(0)) \cdot \epsilon(M_1(0)) \cdot \deg_\epsilon(H_1, \mathcal{O}_1, 0). \tag{4.7}$$

On the other hand, one can show that $t \mapsto [M_t(0)]^{-1}, t \in I$, is a parametrix for $t \mapsto L_t(0), t \in I$. Just using the definitions of parity and orientation it follows that

$$\sigma(L, I) = \epsilon(M_0(0)) \cdot \epsilon(L_0) \cdot \epsilon(M_1(0)) \cdot \epsilon(L_1). \tag{4.8}$$

Since ϵ takes values in $\{\pm 1\}$, (4.6) follows from (4.7) and (4.8).

Observe that without any differentiability assumptions, if one chooses a representation of the homotopy as we just did, then (4.7) holds, and so, in particular, *the absolute value of the degree is always homotopy invariant.*

If $f: X \to X$ is a compact vector field and the point $x_0 \in X$ is a zero of f and is also a regular point of f, i.e., f has an invertible Fréchet derivative at x_0, then the derivative is a linear compact vector field and if $\mathcal{O}$ is a neighborhood of x_0 in which the only zero of f is x_0, then

$$\deg_{L.S.}(f, \mathcal{O}, 0) = \epsilon(f'(x_0)). \tag{4.9}$$

By using the additivity of the Leray-Schauder degree the above extends to the following regular value formula: Let $f: X \to X$ be a C^1 compact vector field and $\mathcal{O}$ be an open bounded subset of X on the boundary of which f does not vanish. If 0 is a regular value of $f: \mathcal{O} \to X$, then

$$\deg_{L.S.}(f, \mathcal{O}, 0) = \sum_{x \in f^{-1}(0) \cap \mathcal{O}} \epsilon(f'(x)). \tag{4.10}$$

The Fréchet derivative of a quasilinear Fredholm map is Fredholm of index 0. But (4.9) does not extend directly. To see why, just consider the case when $\bar{M}: \bar{X} \to \Phi_0(X,Y)$ is a continuous family of isomorphisms, $x_0 \in X$ and $f(x) \equiv M(x)(x - x_0)$ for all $x \in X$. Then the only zero of f is x_0, $f'(x_0) = M(x_0)$ and, by definition, if $\mathcal{O}$ is any neighborhood of x_0 and ϵ is any orientation of $GL(X,Y)$, $\deg_\epsilon(f, \mathcal{O}, 0) = \epsilon(M(0))$. Clearly, $\epsilon(M(0))$ need not be equal to $\epsilon(M(x_0))$. At $x_0 = 0$ there is equality, and, in fact, (4.9) does extend to quasilinear Fredholm mappings in the case that the point x_0 equals 0. Of course, we note that, right at the beginning, the point 0 was selected, more or less arbitrarily, to play a distinguished role in the construction the the degree.

The regular value formula (4.10) has an interesting correspondent in the quasilinear Fredholm degree. It is the following:

A Regular Value Formula *Suppose that $f: \mathcal{O} \to Y$ is quasilinear Fredholm and is also C^1. Suppose also that $f'(0)$ is invertible. Let $\mathcal{O}$ be an open, bounded subset of X*

on whose boundary f does not vanish. If 0 is a regular value of $f: \mathcal{O} \to Y$ and ϵ is an orientation of $GL(X,Y)$, then

$$\deg_\epsilon(f, \mathcal{O}, 0) = \epsilon(f'(0)) \sum_{x \in f^{-1}(0) \cap \mathcal{O}} \epsilon_0(f', x),$$

where

$$\epsilon_0(f', x) = \sigma(f'(tx), [0,1]) \text{ if } f'(x) \in GL(X,Y).$$

This formula suggested an approach to degree theory for smooth Fredholm mappings, in the absence of any assumption of quasilinearity. For C^2-Fredholm mappings defined on simply- connected domains, which are proper on closed, bounded subsets, this construction is outlined in **[8]** and details may be found in **[9]**.

§5 Global Bifurcation for Families Parametrized by R and by S^1

The degree for quasilinear Fredholm mappings described in the last section provides a means of establishing global bifurcation theorems for the zeroes of parametrized families of such maps. As always, to prove such a result, it is first necessary to establish properness.

The properness of linear Fredholm operators on closed, bounded sets and the extendability of the principal part of a quasilinear Fredholm mapping to a parameter space in which X is compactly embedded suffice to verify the following

Lemma 5.1 *Suppose that $f: X \to Y$ is quasilinear Fredholm. Then the restriction of f to each closed, bounded set is proper.*

The following is a convenient way to formulate well-known separation criteria.

Proposition 5.2 *Let $\mathcal{O} \subseteq R \times X$ be open and $f: \mathcal{O} \to Y$ be continuous and such that $f(\lambda, 0) = 0$ if $(\lambda, 0) \in \mathcal{O}$. Assume that the restriction of f to any closed, bounded set is proper. Assume also that $[a,b] \times \{0\} \subseteq \mathcal{O}$,and neither (a,0) nor $(b,0)$ is a bifurcation point of the equation $f(\lambda, x) = 0$. Then either there is global bifurcation of solutions of $f(\lambda, x) = 0$ from $[a,b] \times \{0\}$* or else there is a subset W of $\mathcal{O}$, which is a a bounded neighborhood of $[a,b] \times \{0\}$, and $\eta > 0$ for which the following hold:*

(5.3) $(\lambda, 0) \notin \bar{W}$ *if* $\lambda \leq a - \eta$ *or* $\lambda \geq b + \eta$.

(5.4) $f(\lambda, x) \neq 0$ *if* $0 < \| x \| \leq \eta$ *and* $\lambda \in [a - \eta, a] \cup [b, b + \eta]$.

(5.5) If $(\lambda, x) \in \partial W$ *and* $f(\lambda, x) = 0$, *then* $x = 0$ *and* $\lambda \in [a - \eta, a] \cup [b, b + \eta]$.

Theorem 5.6 [15] *Let $f: R \times X \to Y$ be quasilinear Fredholm and $\mathcal{O} \subseteq R \times X$ be open with $f(\lambda, 0) = 0$ if $(\lambda, 0) \in \mathcal{O}$. Suppose that $[a,b] \times \{0\} \subseteq \mathcal{O}$, that $L_\lambda \equiv D_x f(\lambda, 0)$ exists*

* *Meaning there is a connected set of nontrivial solutions whose closure C intersects $[a,b] \times \{0\}$ and also either C is unbounded, intersects $\partial\mathcal{O}$ or contains a trivial solution not in [a,b].*

as a Fréchet derivative, uniformly in λ, *and depends continuously on* λ, *for* $\lambda \in [a,b] \equiv I$. *Assume that* $L: I \to \Phi_0(X,Y)$ *has invertible end-points and that*

$$\sigma(L,I) = -1.$$

Then there is global bifurcation of the nontrivial solutions of $f(\lambda,x) = 0$ *from* $I \times \{0\}$.

Proof. The assumption that $D_x f(\lambda,0)$ exists as a Fréchet derivative, uniformly in λ, and depends continuously on λ, for $\lambda \in I$, implies that, by possibly shrinking the interval I, we may suppose that $D_x f(\lambda,0)$ exists and is invertible in a neighborhood of each end-point and, furthermore, that neither $(a,0)$ nor $(b,0)$ is a bifurcation point of the equation $f(\lambda,x) = 0$.

We will argue by contradiction. Suppose there is not global bifurcation from $[a,b] \times \{0\}$. According to Lemma 5.1, the restriction of f to any closed, bounded set is proper. Hence, by Proposition 5.2, we may choose a subset W of $\mathcal{O}$ which is a bounded neighborhood of $I \times \{0\}$ and such that (5.3)-(5.5) hold. For $\lambda \in R$, set $W_\lambda = \{x \mid (\lambda,x) \in W\}$ and let f_λ be the restriction of f to W_λ.

Choose ϵ to be an orientation of $GL(X,Y)$. From (5.4) and (5.5) and the homotopy formula for the degree asserted in the previous section it follows that

$$\deg_\epsilon(f_a, W_a, 0) = d \cdot \deg_\epsilon(f_b, W_b, 0), \tag{5.6}$$

where

$$d = \deg_\epsilon(L_a) \cdot \deg_\epsilon(L_b) \cdot \sigma(L,I). \tag{5.7}$$

Since W is bounded, we may choose $\lambda_* < a - \eta$ with $W_{\lambda_*} = \phi$. From the existence property of the degree we conclude that

$$\deg_\epsilon(f_{\lambda_*}, W_{\lambda_*}, 0) = 0. \tag{5.8}$$

Using (5.3), we may select $\delta \in (0,\eta)$ with $\bar{B}(0,\delta) \cap \bar{W}_\lambda = \phi$ if $\lambda \in [\lambda_*, a-\eta]$. But then $W_\lambda = W_\lambda \backslash \bar{B}(0,\delta)$ if $\lambda \in [\lambda_*, a-\eta]$, so that from if $(\lambda,x) \in \partial(W_\lambda \backslash \bar{B}(0,\delta))$ and $\lambda \in [\lambda_*, a]$. Thus, from the homotopy invariance of the absolute value of the degree and (5.8) we obtain

$$\deg_\epsilon(f_a, W_a \backslash \bar{B}(0,\delta), 0) = 0. \tag{5.9}$$

Now, W_a is a neighborhood of 0 and so, by possibly shrinking δ, we may suppose that $\bar{B}(0,\delta) \subseteq W_a$. Thus, by the additive property of degree, (5.9) and the linearization property

$$\deg_\epsilon(f_a, W_a, 0) = \deg_\epsilon(f_a, B(0,\delta), 0) = \epsilon(L_a). \tag{5.10}$$

The same argument implies that

$$\deg_\epsilon(f_b, W_b, 0) = \epsilon(L_b). \tag{5.11}$$

Finally, if we substitute (5.10) and (5.11) into (5.6)-(5.7) we conclude that

$$\sigma(L,I) = 1$$

which is a contradiction. **Q.E.D.**

We may quote Corollary 3.6 in order to obtain from the above theorem the following two criteria for global bifurcation from an isolated singular point of the path of linearizations about the trivial solutions.

Corollary 5.12 *Let $f\colon R \times X \to Y$ be quasilinear Fredholm with $\mathcal{O} \subseteq R \times X$ open and $f(\lambda, 0) = 0$ if $(\lambda, 0) \in \mathcal{O}$. Suppose that $D_x f(\lambda, 0) \in \Phi_0(X, Y)$ exists as a Fréchet derivative, uniformly in λ, and depends continuously on λ for λ in a neighborhood of λ_0. Assume that $(\lambda_0, 0) \in \mathcal{O}$. If $D_x f(\lambda, 0)$ satisfies either assumption (i) or assumption (ii) of Corollary 3.6, then there is global bifurcation of the nontrivial solutions of the equation $f(\lambda, x) = 0$ from $(\lambda_0, 0)$.*

As we discussed in §3, if $GL(X)$ is connected, there are paths $L\colon S^1 \to \Phi_{)}(X, Y)$ having parity -1. Such a path must, of course, have a singular point. Moreover, in view of the fact that the parity of a closed path is a free homotopy invariant of the path, if $K\colon S^1 \to L(X, Y)$ is any path of compact operators, then the path $L + K\colon S^1 \to \Phi_{)}(X, Y)$ must also have a singular point. The homotopy property of the quasilinear Fredholm degree implies that there is the following nonlinear correspondent of this phenomenon where there is data in the "top-order terms of a family"which forces bifurcation to infinity of the zeroes of a family of Fredholm maps parametrized by S^1.

Given a family $f\colon S^1 \times X \to Y$ of quasilinear Fredholm maps, we define the parity of this family $\sigma(f, S^1)$ to be $\sigma(L_0^\lambda, S^1)$, where $f(x) = L_x^\lambda(x) + C(x)$ is any representation of f. Since principal parts of representations differ by compact families, $\sigma(f, S^1)$ does not depend on the choice of representation. Moreover, if $D_x f(\lambda, 0)$ exists as a Fréchet derivative and $D_x f(\lambda, 0)$ depends continuously on $\lambda \in S^1$, then, $\sigma(f, S^1) = \sigma(D_x f(\lambda, 0)$, and, as above, the value of $D_x f(\lambda, 0)$ does not depend on lower-order, i.e., compact, perturbations.

Theorem 5.13 [13] *Suppose that $f\colon S^1 \times X \to Y$ is a family of quasilinear Fredholm maps parametrized by S^1. Assume that there is some parameter value λ_0 at which the set of solutions of the equation $f(\lambda_0, x) = 0$ is bounded and*

$$\deg(f_{\lambda_0}, X, 0) \neq 0.$$

Assume also that

$$\sigma(f, S^1) = -1.$$

Then there is a connected set of solutions of the family of equations $f(\lambda, x) = 0$ which emanates from those at the parameter value $\lambda = \lambda_0$ and which is unbounded.

Proof. We may identify f with a family of quasilinear Fredholm maps parametrized by the interval $[0, 1]$ which has the property that $f_0 = f_1 = f_{\lambda_0}$. Assume that there does not exist a component of solutions having the stated properties. Then using standard separation arguments we may select an open, bounded subset $\mathcal{O}$ of $[0, 1] \times X$ such that $f(\lambda, x) \neq 0$ if $(\lambda, x) \in \mathcal{O}_\lambda \equiv \mathcal{O} \cap (\{\lambda\} \times X)$, and such that $f_{\lambda_0}^{-1}(0) \subset \mathcal{O}_{\lambda_0}$. Certainly , by perturbing by a constant finite-dimensional operator, we may, and will, suppose that $L_a = L_b$ is invertible. According to the homotopy property of the degree, if ϵ is any choice of orientation, then

$$\epsilon(L_a) \cdot \deg_\epsilon(f_{\lambda_0}^{-1}, X, 0)[1 - \sigma(L_0^\lambda, S^1)] = 0.$$

But by the preceding remarks

$$\sigma(L_0^\lambda, S^1) = \sigma(f, S^1)$$

and so we obtain a contradiction. **Q.E.D.**

For a family $f: S^1 \times X \to X$ of compact vector fields, the above theorem has no correspondent, since for such a family the parity $\sigma(f, S^1)$ equals the parity of the family whose constant value is the identity operator and hence the parity is always 1. It should be noted that this is a bifurcation to infinity result without any assumptions at infinity. Moreover, this stands in rather sharp contrast to the classical Leray-Schauder Continuation Principle: the top order terms preclude the existence of any lower order perturbations for which there exist a-priori bounds, uniformly over the whole parameter space, for the set of zeroes.

In [**13**], a general theorem of the above type was established for families parametrized by a manifold and it was used to establish bifurcation to infinity for families of two-point boundary value problems.

REFERENCES

1. A. V. Babin, *Finite dimensionality of the kernel and the cokernel of quasilinear elliptic mappings*, Math. USSR Sb. **22** (1974), 427–454.

2. S. N. Chow and J. K. Hale, *Methods of Bifurcation Theory*, Grundlehren der mathematischen Wissenshaften, vol. 251, Springer-Verlag, New York–Heidelberg–Berlin, 1982.

3. M. G. Crandall and P. H. Rabinowitz, *Bifurcation from simple eigenvalues*, J. Funct. Anal. **8** (1971), 321–340.

4. V. M. Eni, *On the multiplicity of the characteristic values of an operator bundle*, Mat. Issled **4** (1969), 32–41.

5. J. Esquinas, *Optimal multiplicity in local bifurcation theory I: Generalized generic eigenvalues*, J. Diff. Eq. **71** (1988), 206–215.

6. J. Esquinas and J. Lopez-Gomez, *Optimal multiplicity in local bifurcation theory II: General case*, J. Diff. Eq. **75** (1988), 72–92.

7. P. M. Fitzpatrick, *Homotopy, linearization and bifurcation*, Nonlinear Anal. **12** (1988), 171–184.

8. P. M. Fitzpatrick, J. PejsachowiczRabier, *Degré topologique pour les opérateurs de Fredholm non linéaires*, C. R. Acad. Sci. Paris Sér. I Math. **t. 311** (1990), 711–716.

9. ———, *Degree for proper C^2-Fredholm mappings on simply connected domains*, J. Reine Angew. Math. (1992), To Appear.

10. P. M. Fitzpatrick and J. Pejsachowicz, *An extension of the Leray-Schauder degree for fully nonlinear elliptic problems*, in Nonlinear Functional Analysis, F. E. Browder, ed., Proc. Symp. Pure Math., vol. 45 (Part 1), 1986, 425–438.

11. ———, *The fundamental group of the space of linear Fredholm operators and the global analysis of semilinear equations*, Contemporary Math **72** (1988), 47–87.

12. ———, *Local bifurcation for C^1-Fredholm maps*, Proc. Amer. Math. Soc. **105** (1990), 995–1002.

13. ———, *Nonorientability of the index bundle and several-parameter bifurcation*, J. Funct. Anal. **98** (1991), 42–58.

14. ———, *Parity and generalized multiplicity*, Trans. Amer. Math. Soc. **326** (1991), 281–305.

15. ———, *Orientation and the Leray-Schauder Theory for Fully Nonlinear Elliptic Boundary Value Problems*, Mem. Amer. Math. Soc. (1992), In Press.

16. I. Gohberg, S. Goldberg and M. A. Kaashoek, *Classes of Linear Operators Vol 1*, Birkhäuser, 1990.

17. M. Golubitsky and V. Guilleman, *Stable Mappings and Their Singularities*, Springer-Verlag, New York–Heidelberg–Berlin, 1974.

18. E. Hopf, *Abzweigung einer periodischen Lösung von einer Stationären Lösung eines Differentialsystems*, Berichten der Mathematische-Physichen Klasse der Säshsischen Akademie der Wissenschaften zu Leipzig **19** (1942), [Translated into English in The Hopf Bifurcation and its Applications, Marsden and McCracken, Applied Math. Sci. Vol 19, Springer, 1976].

19. J. Ize, *Bifurcation Theory for Fredholm Operators*, Mem. Amer. Math. Soc. **174** (1976).

20. ———, *Necessary and sufficient conditions for multiparameter bifurcation*, Rocky Mountain J. of Math **18** (1988), 305–337.

21. N. Kopell and L. N. Howard, *Bifurcations under nongeneric conditions*, Advances in Math **13** (1974), 274–283.

22. M. A. Krasnosel'skii, *Topological Methods in the Theory of Nonlinear Integral Equations*, MacMillan, 1964.

23. M. A. Krasnosel'skii and P. P. Zabreiko, *Geometrical Methods of Nonlinear Analysis*, Grundleheren der Mathematischen Wissenshaften, vol. 263, Springer-Verlag, New York–Heidelberg–Berlin, 1984.

24. N. H. Kuiper, *The homotopy type of the unitary group of a Hilbert space*, Topology **3** (1965), 19–30.

25. B. Laloux and J. Mawhin, *Multiplicity, the Leray-Schauder formula and bifurcation*, Journal of Diff. Equs. **24** (1977), 309–322.

26. R. J. Magnus, *A generalization of multiplicity and the problem of bifurcation*, Proc. London Math. Soc. **32** (1976), 251–278.

27. B. S. Mitjagin, *The homotopy structure of the linear group of a Banach space*, Uspehki Mat. Nauk. **72** (1970), 63–106.

28. L. Nirenberg, *Topics in Nonlinear Functional Analysis*, Courant Institute Lecture Notes, New York Univ., 1975.

29. J. Pejsachowicz, *K-theoretic methods in bifurcation theory*, Contemporary Math. **72** (1988), 193–205.

30. P. J. Rabier, *Generalized Jordan chains and two bifurcation theorems of Krasnosel'skii*, Nonlinear Anal. **13** (1989), 903–934.

31. P. H. Rabinowitz, *Some global results for nonlinear eigenvalue problems*, J. Funct. Anal. **7** (1971), 487–513.

32. A. I. Šnirel'man, *The degree of a quasi-ruled mapping and a nonlinear Hilbert problem*, Math. USSR-Sb. **18** (1972), 376–396.

33. J. F. Toland, *A Leray-Schauder degree calculation leading to nonstandard global bifurcation results*, Bull. London Math Soc **15** (1983), 149–154.

34. D. Westreich, *Bifurcation at eigenvalues of odd multiplicity*, Proc. Amer. Math Soc. **41** (1973), 609–614.

Continuation Principles and Boundary Value Problems

Mario Martelli

California State University
Fullerton, California 92634

1. Introduction

The basic strategy of a continuation principle is to establish that an equation of the form

$$G(x) = N(x) \tag{1.1}$$

usually defined in a Banach space E, has a solution by embedding (1.1) into a family of equations depending continuously on a parameter λ,

$$G(x) = H(x, \lambda) \tag{1.2}$$

with $H(x, 1) = N(x)$.
An intelligent selection of the family allows the researcher to verify that

- the problem corresponding to $\lambda = 0$ has a solution;
- as λ moves away from zero the corresponding problem is still solvable;
- the solution branch can be controlled so that it will not disappear before $\lambda = 1$.

Consequently, the key ingredients of a continuation principle are

- existence results for the $\lambda = 0$ case;
- permanence properties to insure solvability for λ close to zero;
- a priori bounds to guarantee that the solution branch will persist at least until $\lambda = 1$.

This typical scheme has flexibility for adjustment to situations where different goals are pursued, like establishing the existence of positive eigenvalues and corresponding eigenvectors for nonlinear maps acting on cones of E.

The seminal idea of continuation seems to go back to Poincaré, but are definitely the finite-dimensional Brower and infinite-dimensional Leray-Schauder [26] degree theories which brought the importance of continuation to the forefront of research. In 1961, Granas, A. [22] incorporated it in his theory of essential vector fields, and in the 70's the systematic development of Coincidence Degree Theory by Mawhin, J. [31, 21] and his collaborators enhanced the successful use of continuation methods in solving Boundary Value Problems. Around 1975 Furi, M. - Martelli, M. - Vignoli, A. [12] developed the theory of 0-epi and 0-regular maps, where continuation was combined with fixed point theorems to study problems for which degree theory techniques are either unnecessarily complex or not applicable. The first part of this monograph is based on their ideas, later refined by Furi, M. - Pera, M. P. [13].

Progress on the use of continuation methods in the framework of degree theory has been made by many. Rabinowitz, P. [35], following the footsteps of Krasnosel'skii, M. A., gave some ground breaking applications to Boundary Value Problems and Bifurcation Theory. Fitzpatrick, M. - Pejsachowicz, J. [8, 9] have recently introduced the idea of parity for detecting the real nature of the results established by Rabinowitz and for extending his analysis to a much larger class of problems involving bifurcation for Fredholm operators of index zero. Lately, the attention of researchers has shifted to continuation principles where the a priori bounds component plays a more direct role. Capietto, A.- Mawhin, J. - Zanolin, F. [2, 3] and Furi, M. - Pera, M. P. [18, 19, 20] have given different and interesting applications of this new development.

In these lectures I was tempted to give a general overview of the past history and a complete picture of the present situation. However, I very soon realized that these goals were a bit too ambitious. Therefore I settled for more modest accomplishments and divided my presentation in five parts. The first presents continuation principles in which the key ingredients are some fixed point and extension theorems . The reader will easily recognize in this part the basic ideas developed by Furi, M. - Martelli, M. - Vignoli, A. [12] in their theory of 0-epi and 0-regular maps.

In the second lecture I start the analysis of two typical situations in which continuation plays a determinant role. For historical reasons the two cases are today known as "bifurcation from the parameter line" and "cobifurcation from the state space". I will illustrate how the two apparently different necessary conditions for "starting a branch" are in fact particular cases of a more general principle detected by Furi, M. - Pera, M. P. . A proof of the principle in its full generality will appear in a future paper. In these notes we present a special version of it, which was orally communicated to the author by Furi, M. . Some applications to Boundary Value Problems conclude the lecture.

The third lecture continues on the topic of the second with the analysis of sufficient conditions for "starting the branch" in both the bifurcation and cobifurcation settings, with particular attention to the later. Furi, M. - Pera, M. P. [14] and the author [29] independently developed the ideas presented here. We will show that once again the sufficient conditions used by them in the cobifurcation setting, and the ones used by Krasnosel'skii, M. A. and Rabinowitz, P. in the framework of bifurcation are particular cases of a more general principle, which is presently studied by Furi, M. - Martelli, M. - Pera, M. P. for differentiable manifolds.

In the forth lecture we analyze the so called degenerate case for cobifurcation and the bifurcation from infinity. We also discuss the possibility of extending continuation principles to Fréchet spaces and the difficulties encountered in this enterprise.

Finally, in the fifth lecture, we shall present and compare the recent ideas of Capietto, A. - Mawhin, J. - Zanolin, F. and Furi, M. - Pera, M. P. on the more direct incorporation of a priori bounds in the framework of continuation principles. We will conclude with a list of some open problems and related research projects.

The author would like to thank Furi, M. and Zecca, P. for the invitation and compliment them for the impeccable organization of this C.I.M.E. course.

2. Elementary Continuation Principles and Applications

First, a word of caution. The "elementary" attribute should be taken with a certain care. It is mainly used to underline a prerogative that this expository part have with respect to the others. In fact, here the results are obtained only with an appropriate use of fixed point theorems like the ones of Brower and Schauder, while later we use a more sophisticated approach based on Degree Theory techniques.

As mentioned previously, the basic ideas we introduce in this section are derived from the papers of Furi, M. - Martelli, M. - Vignoli, A. [12] and Furi, M. - Pera, M. P. [14]. The results are not stated in their full generality. Consequently, the proofs are simpler. At a first glance the entire exposition may even seem independent of the papers just mentioned. A closer look however, reveals the strong link between this section and the idea of 0-epi and 0-regular maps introduced in [12] and [14].

We start with a topological result which can be found in Alexander, J. C. [1].

Proposition 2.1 *Let* X *be a compact Hausdorff space, and* K_0, K_1 *be two closed, disjoint subsets of* X. *Assume that no connected component of* X *intersects* K_0 *and* K_1. *Then* X *can be partitioned into two disjoint compact subsets* X_0 *and* X_1, *such that*

$$(2.1) \qquad K_0 \subset X_0, \; K_1 \subset X_1 \; \text{ and } \; X_0 \cup X_1 = X.$$

In the sequel we shall repeatedly use the following two consequences of Proposition 2.1

Corollary 2.1 *Let* E *be a Banach space,* X *be a compact subset of* E. *Assume that* K_0 *and* K_1 *are two disjoint closed subsets of* X *such that no connected component of* X *intersects both. Then there exists an open bounded set* U *of* E *such that*

$$(2.2) \qquad K_0 \subset U, \ \ \bar{U} \cap K_1 = \emptyset, \ \textit{and} \ \partial U \cap X = \emptyset.$$

Proof. According to Proposition 2.1 there is a partition of X into two disjoint closed subsets X_0, and X_1 with $K_0 \subset X_0$ and $K_1 \subset X_1$. Define $d > 0$ by

$$2\,d = \min\{\|x - y\| : x \in X_0, y \in X_1\}$$

and let

$$U = \{z \in E : \|x - z\| < d \text{ for some } x \in X_0\}.$$

It is easy to verify that U satisfies all the properties listed in the Corollary.

Let E be a Banach space and $F = E \times \mathbf{R}$ be endowed with the norm

$$\|(x, \lambda)\| = \max\{\|x\|, |\lambda|\}.$$

Denote by

$$D(n) = \{(x, \lambda) : \|(x, \lambda)\| \le n\}, \qquad S(n) = \{(x, \lambda) : \|(x, \lambda)\| = n\}$$

$$\text{and } B(n) = \{(x, \lambda) : \|(x, \lambda)\| < n\}.$$

Corollary 2.2 *Let* $X_0 \subset E \times \{0\}$ *be compact. Assume that a subset* X *of* $E \times (0, \infty)$ *has the following properties*

a) *The bounded and closed subsets of* $\bar{X}$ *are compact.*

b) $\bar{X} \cap X_0 \neq \emptyset$.

c) *For every* $n \ge 1$ *there exists a non empty connected subset* X_n *of* X *such that* $\bar{X}_n \cap X_0 \neq \emptyset$ *and* $X_n \cap S(n) \neq \emptyset$.

Then there exists an unbounded connected component C *of* X *such that* $\bar{C} \cap X_0 \neq \emptyset$.

Proof. Let G be the standard one-point compactification of F, $G = F \cup \{\infty\}$. The neighborhoods of $\{\infty\}$ are the complements of bounded sets of F. The set $\bar{X} \cup \{\infty\}$ is compact in G. Let $Y_0 = \bar{X} \cap X_0$ and assume that no connected component of $\bar{X} \cup \{\infty\}$ joins Y_0 with $\{\infty\}$. According to Corollary 2.1, there exists an open bounded set U,

$$Y_0 \subset U, \ \infty \notin U \text{ and } \partial U \cap (\bar{X} \cup \{\infty\}) = \emptyset.$$

This however is impossible since for n large the connected set X_n would have to satisfy the two contradictory properties

$$\bar{X}_n \cap X_0 \subset Y_0 \subset U \quad \text{and} \quad (X_n \cap S(n)) \cap U = \emptyset.$$

Hence X_0 and $\{\infty\}$ are joined in $\bar{X} \cup \{\infty\}$ by an unbounded connected component C_∞. The set

$$C = C_\infty \setminus \{\{\infty\} \cup X_0\}$$

has the properties required by the Corollary.

Given two closed, disjoint subsets C_0, C_1 of a metric space X, define $\phi : X \to [0, 1]$ by

$$\phi(x) = \frac{d(x, C_0)}{d(x, C_0) + d(x, C_1)}$$

where $d(x, A) = \inf\{d(x, y) : y \in A\}$. ϕ is continuous and such that

$$\phi(x) = \begin{cases} 0 & \text{iff } x \in C_0 \\ 1 & \text{iff } x \in C_1 \end{cases}$$

We are now ready to establish our first Continuation Principle.

Theorem 2.1 *Let* E *be a Banach space,* $C \subset E$ *be closed convex and* $H : E \times [0,1] \to E$ *be continuous and such that* $\overline{\mathrm{Im}H}$ *is compact. Assume that*

1. *the interior* $\mathring{C}$ *of* C *is not empty;*
2. $H(x, \lambda) \in C$ *for all* $x \in \partial C$ *and* $\lambda \in (0, 1)$.

Let Σ *be the set* $\Sigma = \{(x, \lambda) : H(x, \lambda) = x\}$ *and denote by* $\Sigma_\lambda = \Sigma \cap (E \times \{\lambda\})$ *. Then*

i) Σ_0, Σ_1 *are not empty;*
ii) *there is a connected component* K *of* Σ *which intersects* Σ_0 *and* Σ_1.

<u>Proof</u>. Since $\overline{\mathrm{Im}H}$ is compact and C has interior points we may assume, without loss of generality, that C is bounded and $\mathbf{0}$ is an interior point of C. Then, using a suitable Minkovski functional (see [10]), we renorm E with an equivalent norm $\|.\|_0$ so that C becomes the closed ball $D(1) = D$ centered at $\mathbf{0}$ and with radius 1. Denote by $r : E \to D$ the radial retraction. At this point the proofs that $\Sigma_0 \neq \emptyset$ and $\Sigma_1 \neq \emptyset$ are the same. Therefore we prove only that $\Sigma_0 \neq \emptyset$.

For this purpose, let m be such that $\| H(x, \lambda)\|_0 \leq m$ for all $(x, \lambda) \in D \times [0,1]$ and define

$$T : D(m) \to D(m) \text{ by } T(x) = H(r(x), 0),$$

where $D(m) = \{x \in E : \|x\|_0 \leq m\}$. By the Schauder fixed point theorem there exists $x_0 \in C$ such that $T(x_0) = x_0$. Our assumptions imply $r(x_0) = x_0$. Hence $x_0 = H(x_0, 0)$.

The existence of a connected component K of Σ joining Σ_0 with Σ_1 can be established by contradiction. Assume that no such component exists. Then according to Corollary 2.1 we can find an open bounded set U such that $\Sigma_0 \subset U$, $\overline{U} \cap \Sigma_1 = \emptyset$, and $\partial U \cap \Sigma = \emptyset$. Define

$$\phi(x,\lambda) = \begin{cases} 0 & \text{iff } (x, \lambda) \in \Sigma \cap U^c \\ 1 & \text{iff } (x, \lambda) \in \Sigma \cap U \end{cases}$$

Let m be as above and P be the convex subset of $F = E \times \mathbf{R}$, $P = \{(x, \lambda) : \|x\|_0 \leq m \text{ and } |\lambda| \leq 1\}$. The map $R(x, \lambda) = (H(r(x), \phi(x, |\lambda|)) \ , \phi(x, |\lambda|))$ has a fixed point (x_0, λ_0) .
Therefore $\lambda_0 = \phi(x_0, |\lambda_0|)$, $\lambda_0 \geq 0$ and $H(r(x_0), \phi(x_0, \lambda_0) = x_0$. Our assumptions imply $r(x_0) = x_0$. Consequently, $x_0 = H(x_0, \phi(x_0, \lambda_0)$.

Suppose $(x_0, \phi(x_0, \lambda_0)) \notin U$. Then $\phi(x_0, \lambda_0) = 0$ and $x_0 = H(x_0, 0)$. Hence $(x_0, 0) \in \Sigma_0$, which is contained in U. This contradicts $(x_0, \phi(x_0, \lambda_0)) \notin U$.
Thus $(x_0, \phi(x_0, \lambda_0)) \in U$, and $\phi(x_0, \lambda_0) = 1$. This implies $x_0 = H(x_0, 1)$ and $(x_0, 1) \in \Sigma_1$, which is contained in U^c. We have reached another contradiction. Hence there must be a connected component K of Σ which joins Σ_0 with Σ_1.

The above result can be extended in several directions. For example, by selecting a disk D(q) as a domain of our map H, we can establish a result similar to Theorem 2.1 with much weaker assumptions on H. In particular, using the so-called Kuratowki (see [25]) measure of non compactness we can simply assume that H is a α-*contraction* or, more generally, a *condensing* map. The fixed point theorem of Schauder is substituted by the analogous result of Darbo, G. [4] for α-*contractions* and of Sadovskii, B. N. (see [36] and [27]) for condensing maps.

Extensions to multivalued upper semicontinuous maps are also possible. The fixed point theorems to be used are now due to Kakutani, S. [24] in the compact case when $H(x, \lambda)$ is convex for every (x, λ), and to Eilenberg, S. - Montgomery, D. [6] when, for every (x, λ), $H(x, \lambda)$ is acyclic. The compactness of H can again be substituted with the assumption "H is α-*contractive*" or "H is *condensing*" using basically the same technique of the single-valued case. The introduction of methods based on ideas derived from Algebraic Topology allows further extensions to the so-called admissible multivalued maps which are composition of convex-valued or, more generally, acyclic-valued maps [11]. Further improvements can be achieved by considering maps such that the image of each point never exceeds a finite number p of convex or acyclic components.

We now analyze the structure of the solution set of an equation similar to (1.2), but under somewhat different conditions. We shall assume that the map G is a linear isomorphism L between two Banach spaces E and F, while the domain of the nonlinear operator H is $E\times[0,+\infty)$. The equation we are studying takes the form

$$(2.1) \qquad Lx = H(x, \lambda)$$

The following result holds. The "compactness" assumption should be understood in the sense that H sends bounded sets of $E\times[0, \infty)$ into totally bounded sets of F.

Theorem 2.2 *Let* E, F *be two Banach spaces, and* $L : E \to F$ *be an isomorphism . Assume that a map* $H : E\times[0, \infty) \to F$ *is continuous, compact and such that* $H(x, 0) = \mathbf{0}$ *for all* $x \in E$. *Then there exists and unbounded connected component* C *of*

$$\Sigma = \{(x, \lambda) : Lx = H(x,\lambda), \lambda \geq 0\}$$

such that $(\mathbf{0}, 0) \in C$.

Proof. According to Corollary (2.2) it is enough to show that Σ contains a connected component C_n which starts at $(\mathbf{0}, 0)$ and reaches the set

$$S(n) = \{(x, \lambda) : \|(x, \lambda)\| = n\}, \text{ where } \|(x, \lambda)\| = \max\{\|x\|, |\lambda|\}.$$

Moreover, since L is invertible, we may assume, without loss of generality, that the equalities F=E and L=I hold.

Let us show first that for every positive integer n the equation $x = H(x, \lambda)$ has at least one solution (x_n, λ_n) with $\|(x_n, \lambda_n)\| = n$. In fact, assume that this is not the case for some n.

Notice that $\Sigma \cap D(n)$ is compact and disjoint from S(n). Therefore, we can find a continuous function $\phi : D(n) \to [0,1]$ such that

$$\phi(x, |\lambda|) = \begin{cases} 0 & \text{iff } (x, |\lambda|) \in S(n) \\ 1 & \text{iff } (x, |\lambda|) \in \Sigma\cap D(n) \end{cases}$$

Define $K : D(n) \to D(n)$ by $K(x, \lambda) = (r(H(x, \phi(x, |\lambda|)n)), \phi(x, |\lambda|)n)$, where r is the radial retraction

$$r(x) = \begin{cases} x & \text{if } \|x\| < n \\ \dfrac{nx}{\|x\|} & \text{if } \|x\| \geq n \end{cases}$$

By Schauder's fixed point theorem K has a fixed point (x_0, λ_0). The fixed point does not belong to S(n) for, otherwise,

$$\lambda_0 = \phi(x_0, |\lambda_0|)n = 0 \text{ and } H(x_0, 0) = \mathbf{0},$$

a contradiction. Hence $\|(x_0, \lambda_0)\| < n$. In particular $\|x_0\| < n$. Therefore

$$x_0 = H(x_0, \phi(x_0, |\lambda_0|)n) \text{ and } \lambda_0 = \phi(x_0, |\lambda_0|)n.$$

Hence $\lambda_0 \geq 0$, $(x_0, \lambda_0) \in \Sigma$ and $\phi(x_0, |\lambda_0|) = 1$. Consequently $\lambda_0 = n$, again a contradiction.

At this point the existence of a connected component C_n of Σ which joins $(0, 0)$ with $S(n)$ can be established following essentially the same line of proof used in Theorem 2.1. The only difference is in the definition of the function $R(x, \lambda)$ which now becomes

$$R(x, \lambda) = (r(H(x, \phi(x, |\lambda|)n), \phi(x, |\lambda|)n).$$

Hence we have established that $(0, 0)$ is contained in an unbounded connected component C of Σ.

Theorem 2.2 can also be extended in several directions. In particular, the continuity of H can be replaced by the assumption that H is upper semicontinuous multivalued map with convex or, more generally, acyclic values. The extension to non compact cases requires some care. In many instances the map H take the simpler form $H(x, \lambda) = \lambda N(x)$ and we may assume that "N is a α-contraction with constant p". This modification, however, forces a change in the conclusion of Theorem 2.2. We can prove only the existence of a connected component C of Σ, which contains the point $(0, 0)$, and intersects $S(d)$ for every $d < 1/p$. The branch may disappear as d reaches the value $1/p$. The following example illustrates this observation.

Example 2.1 Let E be the Hilbert space of square summable sequences of real numbers and let D be the subset $D = \{x \in E: \|x\| \leq 1\}$. Define $H : D \times [0, \infty) \to E$ by

$$H(x, \lambda) = H((x_1, x_2, \ldots), \lambda) = \lambda(\sqrt{1 - a^2\|x\|^2}, ax_1, \ldots)$$

with $a < 1$. From $x = H(x, \lambda)$ we derive $\|x\| = \lambda$. Assume $(a\lambda)^2 = 1$. Then $x = 0$ and $\lambda = 0$. Therefore there are no solutions of the equation $x = H(x, \lambda)$ on $S(1/a)$.
For every $\lambda < 1/a$ the only solution of $x = H(x, \lambda)$ is given by

$$x_1(\lambda) = \lambda\sqrt{1 - (a\lambda)^2}, \; x_2(\lambda) = \lambda a x_1, \; x_3 = (\lambda a)^2 x_1, \ldots\ldots$$

Theorem 2.2 can be given a slightly different version for maps acting on cones of Banach spaces. More precisely, we can show that given a cone K in a Banach space E and a continuous, compact map $H : K \times [0, \infty) \to K$, such that $H(x, 0) = 0$, the family of equations

$$(2.2) \qquad x = H(x, \lambda), \quad \lambda \in [0, \infty),$$

admits an unbounded connected component C in K, which contains the point $(0, 0)$.

We present here an application of Theorem 2.1, followed by another of Theorem 2.2. Both applications should be viewed as examples of how the two theorems can be used, and are not intended as illustrations of the full potential of the two results.

Application 2.1 Let $f : [0, 1] \times \mathbf{R} \to \mathbf{R}$ be continuous and such that $f(t, m) > 0$, $f(t,-m) < 0$ for some constant $m > 0$ and all $t \in (0, 1)$. Then the Boundary Value Problem

$$\text{(2.3)} \qquad \begin{cases} x''(t) - f(t, x(t)) = 0 \\ x(0) = x(1) = 0 \end{cases}$$

has a solution.

Consider the Green's function $g : [0,1] \times [0,1] \to \mathbf{R}$

$$\text{(2.4)} \qquad g(t, s) = \begin{cases} t(1 - s) & t \leq s \leq 1 \\ s(1- t) & 0 \leq s \leq t \end{cases}$$

associated to the problem

$$\text{(2.5)} \qquad \begin{cases} x''(t) = y \\ x(0)=x(1) = 0 \end{cases}$$

and rewrite (2.3) as a fixed point problem of a suitable continuous and compact operator T acting on the Banach space $C_0[0, 1]$ of continuous functions $\mathbf{w} : [0, 1] \to \mathbf{R}$ such that $w(0) = w(1) = 0$, endowed with the sup norm $\|w\| = \max\{|w(t)| : t \in [0,1]\}$.

We have

$$\text{(2.6)} \qquad T(x)(t) = \int_0^1 g(t, s)f(s, x(s))ds$$

Embedding (2.6) into a suitable one-parameter family of equations depending on λ, we obtain

$$\text{(2.7)} \qquad H(x, \lambda)(t) = \lambda\int_0^1 g(t, s)f(s, x(s))ds$$

For $\lambda = 0$ we obtain $H(x, 0) = 0$ for all t and the zero function is the only element of Σ_0. According to Theorem 2.1 we have a solution for $\lambda = 1$ if we can find a disk D centered at the origin such that for every $\mathbf{x} \in \partial D$ and $\lambda \in (0, 1)$ we have $H(x, \lambda) \in D$.
Let us choose the disk $D(m)$ and assume that $\mathbf{x} \in S(m)$. Then there exists $t_0 \in (0, 1)$ such that $|x(t_0)| = m$. In the case when $x(t_0) = m$, t_0 is a maximum point, while when $x(t_0) = -m$, t_0 is a minimum point. Hence $x''(t_0) \leq 0$ in the first case, and $x''(t_0) \geq 0$ in the second. Both inequalities contradict the properties of f.

Hence, all assumptions of Theorem 2.1 are satisfied and (2.3) has a solution.

Application 2.2 Let K be a cone in a Banach space, and $T : K \to K$ be a positively homogeneous, order preserving, continuous and compact operator. Assume that there exists a vector **u**, $\|u\| = 1$, and a real number $d > 0$ such that

$$(2.8) \qquad \limsup_{n\to\infty} \frac{\|T^n(u)\|}{d^n} = k > 0.$$

Then there exists a vector $x_0 \in K$, $\|x_0\| = 1$, and a real number $\mu > d$, such that

$$(2.9) \qquad T(x_0) = \mu\, x_0.$$

The proof of this result runs as follows. First we pick any real number $\rho < d$ and we define $T_\rho(x) = \frac{1}{\rho} T(x)$. We can easily verify that the sequence $\{a_n = \|T_\rho{}^n(u)\| : n = 1, 2, ...\}$ is unbounded. Therefore we can find an increasing unbounded subsequence $\{b_q\}$ of $\{a_n\}$ such that the limsup of (2.8) becomes a limit along $\{b_q\}$. We denote by $\{v_q\}$ the subsequence of normalized vectors $\{\frac{T_\rho{}^q(u)}{\|T_\rho{}^q(u)\|}\}$. Notice that $\{v_q\}$ is compact since it can be written as the image, under T, of the bounded sequence $\{\frac{T_\rho{}^{q-1}(u)}{\|T_\rho{}^q(u)\|}\}$. Without loss of generality we can assume that $v_q \to v$, for some $v \in K$, $\|v\| = 1$.

Given a positive integer m we now consider the operator

$$(2.10) \qquad H(m, \rho, x, \lambda) = \lambda(T_\rho(x) + \frac{1}{m} T_\rho(u)).$$

Notice that for $\lambda = 0$ we have $H(m, \rho, x, 0) = \mathbf{0}$. Hence, according to the version of Theorem 2.2 adjusted to operators acting on cones, there exists an unbounded connected component of solutions of the family of equations

$$(2.11) \qquad x = H(m, \rho, x, \lambda)$$

which contains $(\mathbf{0}, 0)$. From $x = \lambda(T_\rho(x) + \frac{1}{m} T_\rho(u))$ we derive

$$(2.12) \qquad x > \lambda T_\rho(x) \quad \text{and} \quad x > \lambda(\frac{1}{m} T_\rho(u))$$

By induction we arrive at

$$(2.13) \qquad x > \lambda^n(\frac{1}{m} T_\rho{}^n(u))$$

We now divide both sides of (2.13) by $\|T_\rho{}^n(u)\|$ and we select those elements of the new inequalities which correspond to the indices q of the convergent subsequence $\{v_q\}$. Rearranging all terms we obtain

$$(2.14) \qquad (\frac{\rho}{\lambda d})^q \frac{d^q}{\|T^q(u)\|} x > \frac{1}{m} v_q$$

We see that $\frac{\rho}{\lambda d}$ cannot be smaller than 1, otherwise, as $q \to \infty$ we would reach the contradiction $v < 0$. Hence $\lambda < \frac{\rho}{d}$. Therefore the unbounded branch must contain a vector $x_\rho(m) \in K$ such that $\|x_\rho(m)\| = 1$ and, correspondingly a value $\lambda_m < \frac{\rho}{d}$. The sequence $\{x_\rho(m)\}$ is obviously compact. Hence there exists $x_0 \in K$, $\|x_0\| = 1$ and a real number $\lambda_0 < \frac{\rho}{d}$ such that

$$(2.15) \qquad x_0 = \lambda_0 T_\rho(x_0)$$

or

$$(2.16) \qquad \mu x_0 = T(x_0)$$

with $\mu = \frac{\rho}{\lambda_0} > d$.

The above result can be extended to α-contractions since we only need that the connected component reaches the boundary of the ball of radius 1. Some details of the proof need to be adjusted. The interested reader should consult [30].

3. A necessary condition for bifurcation

We now turn our attention to a type of problems in which continuation plays a crucial role. From a historical point of view the problems can be divided into two classes. The ones of the first, and older class, have been called bifurcation problems, while the ones of the second class have been labelled as cobifurcation [14] or atypical bifurcation problems [23]. The starting point is an equation of the form

$$(3.1) \qquad Lx = H(x, \lambda)$$

where $L : E \to F$ is a continuous linear operator acting between two Banach spaces E and F, and the map $H : E \times \mathbf{R} \to E$ is continuous and compact.

In the bifurcation case one assumes that L is an isomorphism and $H(0, \lambda) = 0$ for alll $\lambda \in \mathbf{R}$. Hence the pairs $\{(0, \lambda) : \lambda \in \mathbf{R}\}$ are solutions of (3.1). They are called trivial solutions and the goal is to find those values of λ for which a branch of non-trivial solutions (i.e. $x \neq 0$) emanates from the line of trivial solutions $\{0\} \times \mathbf{R}$ and to study the behavior of these branches.

In the cobifurcation case L is a Fredholm operator of index 0, $\dim \mathrm{Ker} L > 0$ and $H(\mathbf{x}, 0) = \mathbf{0}$ for all $\mathbf{x} \in E$. Hence the pairs $\{(\mathbf{x}, 0) : \mathbf{x} \in \mathrm{Ker} L\}$ are solutions of (3.1). As in the previous case, they are called trivial solutions and the goal is to find those vector $\mathbf{x} \in \mathrm{Ker} L$ for which a branch of non-trivial solutions $(\lambda \neq 0)$ emanates from the subspace $\mathrm{Ker} L \times \{0\}$ and to study the behavior of these branches.

Assuming the differentiability of H with respect to $\mathbf{x}$, we can easily verify that, as a consequence of the Implicit Function Theorem, bifurcation can occur from the line of trivial solutions $\{\mathbf{0}\} \times \mathbf{R}$ only at those values of λ for which $\dim \mathrm{Ker}(L - H_{\mathbf{x}}(\mathbf{0}, \lambda)) > 1$.

In the cobifurcation case one can show that a point $(\mathbf{x}_0, 0) \in \mathrm{Ker} L \times \{0\}$ can be the starting point of a non trivial branch only if

$$\text{(3.2)} \qquad \frac{\partial H}{\partial \lambda}(\mathbf{x}_0, 0) \in \mathrm{Im} L$$

In fact, assume that $(\mathbf{x}_0, 0)$ is a cobifurcation point and let $\{(\mathbf{x}_n, \lambda_n) : n=1,...\}$ be such that

$$(\mathbf{x}_n, \lambda_n) \to (\mathbf{x}_0, 0) \text{ and } L\mathbf{x}_n = H(\mathbf{x}_n, \lambda_n).$$

Choose a continuous linear projection $Q : F \to F$ such that $\mathrm{Ker} Q = \mathrm{Im} L$. We obtain

$$\frac{QH(\mathbf{x}_n, \lambda_n) - QH(\mathbf{x}_0, \lambda_n) + QH(\mathbf{x}_0, \lambda_n)}{\lambda_n} = \mathbf{0}.$$

Condition (3.2) will be verified if we show that

$$\lim_{\lambda_n \to 0} \frac{QH(\mathbf{x}_n, \lambda_n) - QH(\mathbf{x}_0, \lambda_n)}{\lambda_n} = \mathbf{0}.$$

Since $H(\mathbf{x}, 0) = \mathbf{0}$ for all $\mathbf{x}$, we obtain

$$QH(\mathbf{x}_n, \lambda_n) - QH(\mathbf{x}_0, \lambda_n) = QH(\mathbf{x}_n, \lambda_n) - QH(\mathbf{x}_n, 0) + QH(\mathbf{x}_n, 0) - QH(\mathbf{x}_0, \lambda_n)$$

Hence

$$\frac{QH(\mathbf{x}_n, \lambda_n) - QH(\mathbf{x}_0, \lambda_n)}{\lambda_n} = \int_0^1 \left[\frac{\partial QH}{\partial \lambda}(\mathbf{x}_n, \tau\lambda_n) - \frac{\partial QH}{\partial \lambda}(\mathbf{x}_0, \tau\lambda_n)\right] d\tau$$

Our task will be completed if we show that

$$\lim_{\lambda_n \to 0} \int_0^1 \left[\frac{\partial QH}{\partial \lambda}(\mathbf{x}_n, \tau\lambda_n) - \frac{\partial QH}{\partial \lambda}(\mathbf{x}_0, \tau\lambda_n)\right] d\tau = \mathbf{0}.$$

The conclusion is true if $\frac{\partial QH}{\partial \lambda}$ is continuous on a neighborhood of $(x_0, 0)$.

At a first glance, it appears that the two necessary conditions for bifurcation and cobifurcation are different. It is our intention to show that this is only an impression. The two conditions represent different forms of the same necessary condition, at least in the case when sufficient regularity can be assumed.

To state this condition in its generality we consider a map $G : X \to Y$, where X and Y are Banach spaces, we denote by $Z = G^{-1}(\mathbf{0})$, and we assume that a finite dimensional C^1 manifold $M \neq \varnothing$ is contained in Z. The manifold M is called the set of trivial solutions of

$$G(x) = \mathbf{0}. \tag{3.3}$$

A point $x_0 \in M$ is called a bifurcation point for equation (3.3) if for every neighborhood U of x_0 we have

$$U \cap (Z \setminus M) \neq \varnothing. \tag{3.4}$$

In the following theorem, we denote by $T(x, M)$ the tangent space to M at the point $x \in M$. The result should be credited to Furi, M. - Pera, M. P. and will appear in a forthcoming paper of Furi, M.- Martelli, M. - Pera, M. P.. The proof presented here has been orally communicated to the author by Furi, M.

Theorem 3.1 *Let* $G = L - K$, *where* L *is a bounded, linear Fredholm operator of index* 0 *and* K *is continuous, compact and of class* C^1. *Then* $x_0 \in M$ *is a bifurcation point for the equation* $G(x)=\mathbf{0}$ *only if*

$$T(x_0, M) \subset \text{Ker}(L - K_x(x_0)) \text{ and } \text{Ker}(L - K_x(x_0)) \neq T(x_0, M).$$

Proof. First we show that $T(x_0, M)$ is included in $\text{Ker}(L - K_x(x_0))$. This follows as an easy consequence of the fact that the composition $G \circ j$ of the inclusion map $j : M \to X$ and of G is identically zero. Thus the restriction of the derivative $L - K_x(x_0)$ to $T(x_0, M)$ is the zero linear operator. Consequently,

$$T(x_0, M) \subset \text{Ker}(L - K_x(x_0)).$$

Notice that this is true for every point $x_0 \in M$.

At this point, noticing that the bifurcation problem is a local problem at the point x_0 and that M is a differentiable manifold, we can assume, without loss of generality, that there exist a finite dimensional space E_m, $x_0 \in E_m$, and a positive real number r, such that

$$B(x_0, \rho) \cap E_m \text{ is contained in Z and}$$

$$B(x_0, \rho) \cap (Z \setminus E_m) \neq \emptyset$$

for all $\rho \leq r$. We then write $\mathbf{x} = \mathbf{u} + \mathbf{v}$, with $\mathbf{u} \in E_m$, $\mathbf{v} \in E_v$, where E_v is a direct summand of E_m. Pick a sequence $\{x_n\}$,

$$x_n \in Z,\ x_n \to x_0, \text{ with } v_n = x_n - u_n \neq \mathbf{0} \text{ and } u_n \in B(x_0, r) \cap E_m \text{ for all n.}$$

Then

$$\mathbf{0} = G(x_n) - G(u_n) = \int_0^1 G_x(x_n + t(u_n - x_n))(x_n - u_n)\, dt \tag{3.5}$$

From (3.5) we derive

$$L \frac{v_n}{\|v_n\|} = K_x(x_0) \frac{v_n}{\|v_n\|} - \int_0^1 [K_x(x_n + t(u_n - x_n)) - K_x(x_0)](\frac{v_n}{\|v_n\|})\, dt$$

We can assume that the sequence $\{K_x(x_0) \frac{v_n}{\|v_n\|}\}$ is convergent, while, as $n \to \infty$, the integral term goes to $\mathbf{0}$. Hence $L\frac{v_n}{\|v_n\|}$ converges. Since L is a bounded Fredholm operator of index 0, we obtain, using possibly a suitable subsequence, that $\{\frac{v_n}{\|v_n\|}\}$ converges to some vector $\overline{v}$ of norm 1, $\|\overline{v}\| = 1$, $\overline{v} \in E_v$, such that

$$L\overline{v} - K_x(x_0)\overline{v} = \mathbf{0}.$$

Hence $\mathrm{Ker}(L - K_x(x_0))$ contains, but does not coincide with $T(x_0, M)$.

Let us see the form taken by the condition expressed by Theorem 3.1 in the case of the classical bifurcation from the line of trivial solutions $\{\mathbf{0}\} \times \mathbf{R}$ or cobifurcation from the subspace $\mathrm{KerL} \times \{0\}$.

1. <u>The necessary condition for bifurcation from</u> $(\mathbf{0}, \lambda_0)$.

The operator $G : E \times \mathbf{R} \to E$ is $G(y) = Lx - H(x, \lambda)$, with $y = (x, \lambda)$. Since L is an isomorphism and H is compact, G is a Fredholm operator of index 0. From $H_\lambda(\mathbf{0}, \lambda_0) = \mathbf{0}$ we see that the line of trivial solutions is included in $\mathrm{KerG}_y(\mathbf{0}, \lambda_0)$. The necessary condition requires the existence of a vector $v \in E$, $\|v\| = 1$ such that

$$Lv - H_x(\mathbf{0}, \lambda_0)v = \mathbf{0}, \text{ i.e. } \dim\mathrm{Ker}(L - H_x(\mathbf{0}, \lambda)) > 1,$$

as previously found.

2. <u>The necessary condition for cobifurcation from</u> $(\mathbf{x}_0, 0)$, $\mathbf{x}_0 \in$ KerL.

The operator G is the same as above. Since $H_{\mathbf{x}}(\mathbf{x}_0, 0) = 0$, we derive the inclusion of the subspace $\text{KerL} \times \{0\}$ in $\text{KerG}_{\mathbf{y}}(\mathbf{x}_0, 0)$. The necessary condition requires the existence of a vector $\mathbf{x}_0 \in$ KerL such that

$$L\mathbf{v} - \mu H_{\lambda}(\mathbf{x}_0, 0) = \mathbf{0},$$

for pairs $(\mathbf{v}, \mu)$ with $\mathbf{v} \notin$ KerL.
This in turn requires $H_{\lambda}(\mathbf{x}_0, 0) \in$ ImL, as previously found.

In the following examples we illustrate the form taken by the necessary condition in some periodic boundary value problems.

Example 3.1 Consider the periodic boundary value problem

$$\begin{cases} x'' + x & = & \lambda(x^3 + f(t)) \\ x(0) - x(2\pi) & = & 0 \\ x'(0) - x'(2\pi) & = & 0 \end{cases}$$

where f is a continuous function, periodic of period 2π and such that $a_1{}^2 + b_1{}^2 > 0$, with a_1 and b_1 the coefficients of Cos t and Sin t in the Fourier expansion of f.

The elements of our Banach space E are those periodic functions **x** of period 2π, which are continuous together with their first and second derivative. The norm in E is given by

$$\|\mathbf{x}\| = \max_{0 \le t \le 2\pi} |x(t)| + \max_{0 \le t \le 2\pi} |x'(t)| + \max_{0 \le t \le \pi} |x''(t)|$$

The elements of the Banach space F are the continuous periodic functions **y** of period 2π . The norm in F is given by

$$\|\mathbf{y}\| = \max_{0 \le t \le 2\pi} |y(t)| \ .$$

The action of the linear, continuous operator $L : E \to F$ is $(L\mathbf{x})(t) = x''(t) + x(t)$. KerL is given by all linear combinations of Cos t and Sin t. Moreover $\mathbf{y} \in$ ImL if and only if

$$\int_0^{2\pi} y(t)\text{Cos}\, t \, dt = \int_0^{2\pi} y(t)\text{Sin}\, t \, dt = 0$$

Thus L is a continuous, linear, Fredholm operator of index 0. The nonlinear map $H : E \times \mathbf{R} \to F$ is defined by $H(\mathbf{z}, \lambda) = \lambda([z(t)]^3 + f(t))$. Since, by Ascoli-Arzela theorem, the embedding of E into F is compact we can consider H as a continuous and compact operator from $E \times \mathbf{R}$ into F.

We see that for $\lambda = 0$ all elements of KerL are solutions of our problem. The search for those functions x_0 in KerL such that the pair $(x_0, 0)$ satisfies the necessary condition for being the starting point of a branch of non trivial solutions leads us to those vectors $\mathbf{a} = (a, b) \in \mathbf{R}^2$ such that

$$\int_0^{2\pi}\{[a\mathrm{Cos}\,t + b\mathrm{Sin}\,t]^3 + f(t)\}\mathrm{Cos}\,t\,dt = \int_0^{2\pi}\{[a\mathrm{Cos}\,t + b\mathrm{Sin}\,t]^3 + f(t)\}\mathrm{Sin}\,t\,dt = 0.$$

These equalities produce the system

$$\begin{cases} a^3 + ab^2 &= -\frac{4}{3}a_1 \\ a^2b + b^3 &= -\frac{4}{3}b_1 \end{cases}$$

Our assumption on a_1 and b_1 implies that the above system has only one solution given by

$$\bar{a} = \frac{-\frac{4}{3}a_1}{\sqrt[3]{a_1^2 + b_1^2}}, \quad \bar{b} = \frac{-\frac{4}{3}b_1}{\sqrt[3]{a_1^2 + b_1^2}}$$

Therefore only the pair $(\bar{a}\,\mathrm{Cos}\,t + \bar{b}\,\mathrm{Sin}\,t, 0)$ can be the starting point of a non trivial branch of solutions of our periodic boundary value problem.

In the next example, we show how a different position of the real parameter λ can change the bifurcation problem.

Example 3.2 Let

$$\begin{cases} x'' &= \lambda(-x + x^3 + f(t)) \\ x(0) - x(2\pi) &= 0 \\ x'(0) - x'(2\pi) &= 0 \end{cases}$$

Our Banach spaces E and F are the same as in the previous example. The action of the continuous linear operator L is now $(Lx)(t) = x''(t)$. The kernel of L are the constant functions, and its image are those continuous functions $y \in F$ such that

$$\bar{y} = \int_0^{2\pi} y(t)dt = 0.$$

Hence L is a continuous linear, Fredholm operator of index 0. The nonlinear operator H is defined in a manner analogous to the previous example, and is continuous and compact. The search for pairs of the form $(r, 0)$ as starting points for a branch of non trivial solutions leads us to the equation

$$-r + r^3 + \bar{f} = 0, \text{ with } \bar{f} = \frac{1}{2\pi}\int_0^{2\pi} f(t)dt$$

We see that the possible starting points are

$$\text{three if } -\frac{2}{3\sqrt{3}} < \bar{f} < \frac{2}{3\sqrt{3}}$$

$$\text{one if } |\bar{f}| > \frac{2}{3\sqrt{3}}$$

$$\text{two if } |\bar{f}| = \frac{2}{3\sqrt{3}}$$

It is legitimate to ask if the necessary condition is also sufficient. The following example shows that in the cobifurcating situation this is not true.

Example 3.2 Let $L : \mathbf{R}^2 \to \mathbf{R}^2$, and $H : \mathbf{R}^2 \times \mathbf{R} \to \mathbf{R}^2$, be such that

$$Lx = L(x, y) = (0, x), \quad H(x, y, \lambda) = \lambda(x^2 + y^2, x^2 + y^2 - x+1).$$

Then $\text{KerL} = \{(x, y) : x = 0\}$ and our necessary condition requires that $(y^2, y^2 + 1)$ belongs to ImL. This implies $y = 0$. From $(0, x) = \lambda(x^2 + y^2, x^2 + y^2 - x + 1)$ with $\lambda \neq 0$ we derive $x^2 + y^2 = 0$ or $x = y = 0$. Then $(0, 0) = \lambda(0, 1)$ requires $\lambda = 0$. Hence the only possible pair $(\mathbf{0}, 0)$ is not a starting point of a branch of non trivial solutions.

We conclude this part by showing, with an additional example, that the necessary condition is not sufficient also in the bifurcating situation. The example is taken from notes by Berestycki, H. of lectures given by Rabinowitz, P. at the Université Paris VI in 1975.

Example 3.3 Let $E = \mathbf{R}^2 = F$, $L = I$ and $H(\mathbf{x}, \lambda) = \lambda(x, y) + (y^3, -x^3)$. Then we have $H(\mathbf{0}, \lambda) = \mathbf{0}$ for all λ and the necessary condition is satisfied by the pair $(\mathbf{0}, 1)$. The equality $(x, y) = H(\mathbf{x}, \lambda)$ implies

$$(1 - \lambda)x = y^3 \text{ and } (1-\lambda)y = -x^3$$

For $\lambda = 1$ one gets $x = y = 0$. For $\lambda \neq 1$ one arrives at $x^4 + y^4 = 0$. Hence, once again, $x = y = 0$. Thus the pair $(\mathbf{0}, 1)$ is not the starting point of a branch of non trivial solutions of $\mathbf{x} = H(\mathbf{x}, \lambda)$.

Therefore, the necessary condition for bifurcation stated in Theorem 3.1 is not sufficient. In the next part we shall present a sufficient condition in the case of cobifurcation and try to see how this condition compares with the one given by Rabinowitz in the case of bifurcation.

4. A sufficient condition for cobifurcation

In this section we present a sufficient condition for cobifurcation and we study the behavior of the branch of non trivial solutions emanating from the pair $(x_0, 0)$, with $x_0 \in \mathrm{KerL}$. The sufficient condition given by Rabinowitz for bifurcation from the line of trivial solutions at the point $(\mathbf{0}, \lambda_0)$ and the one provided for cobifurcation seem to be once more particular cases of a general result which gives a sufficient condition for a point of the manifold M to be a bifurcation point for the equation $G(x) = \mathbf{0}$. Furi, M., Martelli, M. and Pera, M. P. are presently investigating this problem and the results of their research will appear in a forthcoming paper.

The theory presented here is essentially contained in [29], [23] and [14]. Furi, M. and Pera, M. P. [14] used the name cobifurcation for the first time. They obtained a theorem analogous to Theorem 4.1 below, which is essentially due to Martelli, M. [29]. They reached the result independently of Martelli and at the same time.

Their approach is based on a global behavior of the nonlinear term on $E \times \{0\}$, rather than in its local behavior around a starting point. The conclusions reached here are somehow parallel to the ones established by Rabinowitz in the case of bifurcation and this parallelism will be underlined at the appropriate time.

With reference to the previous section, we consider the equation

$$Lx = H(x, \lambda). \tag{4.1}$$

We assume that L is a linear, bounded, Fredholm operator of index 0, acting between two Banach spaces E and F, $H : E \times \mathbf{R} \to F$ is continuous, compact and such that $H(x, 0) = \mathbf{0}$ for all $x \in E$. We also assume that a point $(x_0, 0)$, $x_0 \in \mathrm{KerL}$, satisfies the necessary condition mentioned in Theorem 3.1. Observe that we can consider the vector field $V : \mathrm{KerL} \to \mathrm{ImQ}$ defined by

$$V(x) = \frac{\partial QH}{\partial \lambda}(x, 0). \tag{4.2}$$

V can be identified with a vector field from $\mathbf{R}^q$ into itself, where q is the dimension of KerL. The zeroes of V provide the first component of those pairs which are the possible points of depart for branches of non trivial solutions of (4.1). We want to show now that a sufficient condition is given by an additional property of the vector field V at those particular points. This condition is contained in the following theorem, in which we denote by Σ the set of non trivial solutions of (4.1). Thus

$$\Sigma = \{(x, \lambda) : Lx = H(x, \lambda), \lambda \neq 0\},$$

and

$$\Sigma_+ = \{(x, \lambda) \in \Sigma : \lambda > 0\}, \Sigma_- = \{(x, \lambda) \in \Sigma : \lambda < 0\}.$$

Theorem 4.1 *Assume that a point* $(x_0, 0)$, $x_0 \in \text{KerL}$ *satisfies the necessary condition and the additional condition*

$$(4.3) \qquad \det(V_x(x_0)) \neq 0$$

Then there are two connected branches C_- *and* C_+ *of* Σ_- *and* Σ_+ *respectively, such that*

i) *the closure of each branch contains the point* $(x_0, 0)$,

ii) *each branch is either unbounded or its closure contains another point* $(y_0, 0)$, $y_0 \in \text{KerL}$, $y_0 \neq x_0$.

Before proving Theorem 4.1 we would like to mention that the existence of a connected branch for small values of λ can be established using the Implicit Function Theorem. This is the approach of Prodi, G. - Ambrosetti, A. in [34]. To establish the behavior of the branch, as mentioned in the second part of Theorem 4.1, it seems that more than the Implicit Function Theorem is needed and we shall use a degree theory approach. We assume that the reader has some familiarity with it, and we simply recall here the fundamental properties which will be needed in the sequel.

Let E, F be Banach spaces, $A : E \to F$ be an isomorphism and $N : E \to F$ be continuous and compact. Let U be an open and bounded set of E, and assume that the equation

$$(4.4) \qquad Ax = N(x)$$

does not have any solution on the boundary of U.
Then we can find an integer-valued function $\deg(A, N, U)$ with the following properties.

1. Solvability : $\deg(A, N, U) \neq 0$ implies that (4.4) has a solution $x \in U$.

2. Additivity : if $U = U_1 \cup U_2$, with $U_1 \cap U_2 = \emptyset$, then

$$\deg(A, N, U) = \deg(A, N, U_1) + \deg(A, N, U_2)$$

3. Excision : if C is a closed subset of U and $Ax \neq N(x)$ for all $x \in C$ then

$$\deg(A, N, U) = \deg(A, N, U \backslash C)$$

4. Homotopy: let $N : E \times [a, b] \to F$, and $U \subset E \times [a, b]$ be open and bounded . Define $U_\lambda = (E \times \{\lambda\}) \cap U$ and assume that $Ax \neq N(x, \lambda)$ for all $(x, \lambda) \in \partial U$.
Then $\deg(A, N(., \lambda), U_\lambda)$ is independent of λ. In particular

$$\deg(A, N(., a), U_a) = \deg(A, N(., b), U_b)$$

5. Finite dimensional reduction: assume that ImN is contained in a finite dimensional subspace F^m of F. Then, denoting by A_m and N_m the restrictions of A and N to $A^{-1}(F^m)$, we have

$$\deg(A, N, U) = \deg(A_m, N_m, U \cap A^{-1}(F^m))$$

With these properties at hand we are ready to prove Theorem 4.1.

Proof. The proof is done for C_+. Only minor changes are needed to adjust it to C_-. Consider the operator $A : E \to F$ defined by $Ax = Lx + V_x(x_0)$. It can be easily seen that A is an isomorphism. Set

$$N(x, \lambda) = \begin{cases} V_x(x_0)(x - x_0) - \dfrac{QH(x, \lambda)}{\lambda} + (I - Q)H(x, \lambda) & \lambda > 0 \\ V_x(x_0)(x - x_0) - V(x) & \lambda = 0 \end{cases}$$

The properties of H and the definition of V imply that N is continuous and compact. Notice that for $\lambda > 0$ the solution sets of the equations

$$Lx = H(x, \lambda) \text{ and } Ax = N(x, \lambda)$$

coincide, while for $\lambda = 0$ the solutions of the second equations are only the pairs which satisfy the necessary condition.

Assumption (4.3) implies the existence of a positive real number r such that $V(x) \neq 0$ for all $x \in \text{KerL}$, $0 < \|x - x_0\| \leq r$. Thus $\deg(A, N(., 0), B(x_0, r))$ is defined.

The image of $N(., 0)$ is contained in the finite dimensional space ImQ. We can apply the finite dimensional reduction property to obtain

$$\deg(A, N(., 0), B(x_0, r)) = \deg(A, N(., 0), B(x_0, r) \cap \text{KerL}).$$

Since L vanishes on $B(x_0, r) \cap \text{KerL}$ and $V_x(x_0)$ cancels out, the degree is computed simply using the vector field V. Assumption (4.3) implies that the degree is different from 0.

The properties of A and N imply the existence of a positive real number d such that $\lambda \leq d$ implies $Ax \neq N(x, \lambda)$ if $\|x - x_0\| = r$. Hence, in the cylinder $O = B(x_0, r) \times [0, d]$ we have, by the homotopy property, $\deg(A, N(., \lambda), O_\lambda) \neq 0$. Therefore, by the solvability property, the equation $Lx = H(x, \lambda)$ is solvable inside the cylinder O for every value of the parameter λ.

Denote by S these solutions and set $S_\lambda = S \cap O_\lambda$. Assume that $S_0 = (x_0, 0)$ and S_d are not connected in S. Since S is compact, using Proposition 2.1 we can find an open set U contained in O such that

$$S_0 \subset U, \ \bar{U} \cap S_d = \emptyset, \text{ and } \partial U \cap S = \emptyset.$$

Using the excision property we can show that

$$\deg(A, N(., 0), U_0) = \deg(A, N(., 0), O_0) \neq 0.$$

Hence $\deg(A, N(., d), U_d) \neq 0$. However, no solutions of $Lx = H(x, \lambda)$ can be found in U_d. This contradiction shows that there exists a connected component G_S of S which joins $(x_0, 0)$ with S_d.

Denote by G_+ the connected component of the closure of Σ_+ which contains G_S and assume that G_+ is bounded and does not contain any other point of the form $(y_0, 0)$, $y_0 \in \operatorname{KerL}$. We can find three positive real numbers ρ ($\rho > r$), δ and σ such that G_+ is contained in the interior of the closed cylinder $K = D(x_0, \rho) \times [0, \delta]$. Moreover $\lambda \leq \sigma$ and $(x, \lambda) \in G_+$ imply the inequality $2 \|x - x_0\| < r$. Let us denote by Σ_K the points of K which are in the closure of Σ_+ and by G_b the points (x, λ) of Σ_K which have one of the following properties:

$$1) \|x\| = \rho, \quad 2) \lambda = \delta, \quad 3) \lambda = 0 \text{ and } \|x - x_0\| \geq r.$$

Since G_+ and G_b are closed and not connected in the compact set Σ_K we can find an open set W contained in the cylinder K such that

$$G_+ \subset W, \ \overline{W} \cap G_b = \emptyset, \text{ and } \partial W \cap \Sigma_K = \emptyset.$$

It is easily seen that $\deg(A, N(., 0), W_0) = \deg(A, N(., 0), O_0) \neq 0$. Moreover, there exists $\lambda < \delta$ such that $\Sigma_K \cap W_\lambda = \emptyset$. Hence $\deg(A, N(., 0), W_\lambda) = 0$, contradicting the homotopy property.

Consequently, we can find a connected component C_+ of Σ_+ such that its closure contains $(x_0, 0)$ and C_+ is either unbounded or its closure contains at least one more point of the form $(y_0, 0)$ with $y_0 \in \operatorname{KerL}$, $y_0 \neq x_0$.

The following examples show that both situations outlined at the end of Theorem 4.1 are possible.

Example 4.1 Let $E = F = \mathbf{R}^2$, $L : E \to F$, $H : E \times \mathbf{R} \to F$ be defined by

$$Lx = L(x, y) = (x, 0), \ H(x, y, \lambda) = \lambda(1, \lambda x + y^2 - y).$$

We have

$$\operatorname{KerL} = \{(0, y) : y \in \mathbf{R}\}, \ Q(x, y) = (0, y)$$

and

$$QH(x, y, \lambda) = \lambda(0, \lambda x + y^2 - y).$$

Therefore $\frac{\partial QN}{\partial\lambda}(0, y, 0) = (0, y^2 - y) \in \mathrm{ImL}$ if and only if either $y = 1$ or $y = 0$. Hence the possible cobifurcation points are $(0, 0, 0) = (x_0, 0)$ and $(0, 1, 0) = (x_1, 0)$. Since KerL and ImQ are one dimensional the map V is a real function of a real variable, $V'(0) = -1$, $V'(1) = 1$. Thus both points satisfy the sufficient condition.

For $\lambda \neq 0$ we have $Lx = H(x, \lambda)$ if and only if $x = \lambda$ and $0 = x^2 + y^2 - y$. Consequently the closure of Σ is the ellipse obtained as intersection of the circular cylinder $x^2 + y^2 - y = 0$ with the plane $x = \lambda$. The closure of C_+ is an arc of this ellipse, and the closure of C_- is the other arc. Both branches start from KerL and go back to it.

Example 4.2 Let $E = F = \mathbf{R}^2$, $L : E \to F$, $H : E \times \mathbf{R} \to F$ be defined as follows

$$L(x, y) = (0, x), \qquad H(x, y, \lambda) = \lambda(y - 1, 1).$$

Then

$$\mathrm{KerL} = \{(0, y) : y \in \mathbf{R}\},\ Q(x, y) = (x, 0) \text{ and } QH(x, y, \lambda) = \lambda(y - 1, 0).$$

Therefore, we have

$$\frac{\partial QN}{\partial\lambda}(0, y, 0) = (y - 1, 0).$$

The only possible starting point is the vector $(0, 1, 0)$. Since the sufficient condition is satisfied, both C_+ and C_- are unbounded and $(0, 1, 0)$ is contained in their closure. The branches are given by the intersection of the two planes $x = \lambda$ and $y = 1$.

Remark 4.1 In some cases the condition expressed in (4.2) needs to be adjusted in a suitable manner, by replacing it with

$$\text{(4.2a)} \qquad V(x) = \frac{\partial^i QH}{\partial^i\lambda}(x, 0)$$

The easiest way in which such situation arises is when $H(x, \lambda) = \lambda^i N(x)$. The proof of the theorem remains unchanged.

The following Corollary is essentially Theorem 2.2, which can obviously be derived as a consequence of Theorem 4.1

Corollary 4.1 (see [35]) *Let* $H : E \times \mathbf{R} \to E$ *be continuous, compact and such that* $H(x, 0) = 0$ *for all* $x \in E$. *Then the equation*

$$\text{(4.5)} \qquad x = H(x, \lambda)$$

admits two unbounded connected components

$$C_+, C_+ \subset E\times[0,+\infty) \quad and \quad C_-, C_- \subset E\times(-\infty,0],$$

such that $(\mathbf{0}, 0) = C_+ \cap C_-$.

Corollary 4.2 below can be established by following the guidelines of the proof of Theorem 4.1 from the point where the sufficient condition starts playing its role and the search for non trivial branches of solutions of the equation $\mathbf{Lx} = \mathbf{H}(\mathbf{x}, \lambda)$ is reduced to the search of non trivial branches of solutions of the equation $\mathbf{Ax} = \mathbf{N}(\mathbf{x}, \lambda)$.

Corollary 4.2 *Let* $A : E \to F$ *be an isomorphism and* $N : E \times \mathbf{R} \to F$ *be of class* C^1 *and compact. Assume that* $A\mathbf{x}_0 - H(\mathbf{x}_0, 0) = \mathbf{0}$ *and* $A - H_\mathbf{x}(\mathbf{x}_0, 0)$ *is an isomorphism. Then the point* $(\mathbf{x}_0, 0)$ *belongs to the closure of a connected component* C_+ *of solutions of*

$$(4.6) \qquad \mathbf{Ax} = \mathbf{N}(\mathbf{x}, \lambda) \qquad \lambda > 0$$

such that

- *either* C_+ *is unbounded*
- *or its closure contains another point* $(\mathbf{y}_0, 0)$ *such that*

$$A\mathbf{y}_0 - H(\mathbf{y}_0, 0) = \mathbf{0}.$$

At this point the interested reader should compare Theorem 4.1 with the result obtained by Rabinowitz, P. [35] in the case of bifurcation from the line of trivial solutions. He studies an equation of the form

$$(4.7) \qquad \mathbf{x} = \lambda \mathbf{Lx} + \mathbf{R}(\mathbf{x}, \lambda)$$

where B is a compact linear operator and R is continuous, compact and such that

$$(4.8) \qquad \lim_{\mathbf{x}\to\mathbf{0}} \|\mathbf{x}\|^{-1}\mathbf{R}(\mathbf{x}, \lambda) = \mathbf{0}$$

uniformly on bounded intervals of λ. He proves the following result.

Theorem 4.2 (see [35]) *Assume that* λ_0 *is a characteristic value of* L *of odd multiplicity. Then there is a connected component* C *of* Σ, *the set of non trivial solutions of* (4.7) $[\mathbf{x} \neq \mathbf{0}]$ *such that*

- $(0, \lambda_0) \in \overline{C}$, *the closure of* C.
- *Either* C *is unbounded or* $\overline{C}$ *contains another point of the form* $(\mathbf{0}, \lambda_1)$, *where* $\lambda_1 \neq \lambda_0$ *is a characteristic value of* B.

The similarity between the behavior of the connected components of Σ in Theorems 4.1 and 4.2 is evident. In both cases the components are either unbounded or their closure contains another point of the same type as the starting point. Actually this last statement can be made more precise. The observation we are going to present deserves a remark.

Remark 4.2 In both cases, whenever the connected component goes back either to the line of trivial solutions or to KerL, we can prove something more precise. In the cobifurcation case we can establish that the determinant of the vector field V must have opposite sign at the two points. In the bifurcation case it can be shown that the index changes sign an even number of times between the two values of λ. This remark will be proved in the forthcoming paper of Furi, M. - Martelli, M. - Pera, M. P. .

The two results have other similarities. We already mentioned the necessary condition which is essentially the same in both cases. As far as the sufficient conditions is concerned, they seem different, at least at a superficial glance . This impression is however not correct . In the same paper of Furi, M.- Martelli, M. - Pera, M. P. it will be shown that in both the bifurcation and the cobifurcation cases one can define a vector field V such that the zeroes of V are the possible starting points of the branches of non trivial solutions (necessary condition) and an additional property satisfied by V constitutes the sufficient condition for a zero of V to actually be the origin of a branch.

Both results are also based on two similar technical lemmas. For the cobifurcation case we did not state the lemma explicitly, but its content is incorporated in the proof of Theorem 4.1 . Its explicit formulation is the following.

Lemma 4.1 *Assume that* $(x_0, 0)$ *satisfies the necessary and sufficient condition for cobifurcation and that the connected branch* C_+ *of non trivial solutions is bounded and its closure contains only the point* $(x_0, 0)$. *Then there exists an open and bounded set* U *and two positive real numbers* ε, d *such that*

1. $C \subset U$, $\partial U \cap \Sigma_+ = \emptyset$;

2. $\dfrac{\partial QH}{\partial \lambda}(x, 0) \neq 0$ *for all* $x \in \mathrm{KerL}$, $0 < \| x - x_0 \| < \varepsilon$

3. $U \cap (E \times [0, d]) = B(x_0, \varepsilon) \times [0, d]$.

The technical lemma used by Rabinowitz to establish his result is similar. Its precise formulation is as follows.

Lemma 4.2 *Assume that* λ_0 *is a characteristic value of* L *and let* C *be the connected component of non trivial solutions whose closure contains* $(0, \lambda_0)$. *If* C *is bounded and its closure does not contain any other point of the form* $(0, \lambda_1)$ *with* λ_1 *another characteristic value of* L, *then we can find an open and bounded set* O *and two positive real numbers* ε, d *such that*

1. $C \subset O$, $\partial O \cap \Sigma_+ = \emptyset$;

2. *every* $\lambda \in (\lambda_0 - \varepsilon, \lambda_0 + \varepsilon)$ *is not a characteristic value of* L;

3. $U \cap (B(\mathbf{0}, d) \times \mathbf{R}) = B(\mathbf{0}, d) \times (\lambda_0 - \varepsilon, \lambda_0 + \varepsilon)$.

The observations made so far reveal that bifurcation and cobifurcation are two particular cases of a more general phenomenon involving operator equations. The space of trivial solutions is different, but the points in common between the two cases are overwhelming. The general theory to be presented in the paper of Furi - Martelli- Pera will further underline the similarities between the two cases, which should be obtained as corollaries of a result in which the two flat manifolds of trivial solutions will be replaced by a differentiable finite dimensional manifold $M \subset G^{-1}(\mathbf{0}) = X$. We shall see in the next section that there are still issues to be resolved, particularly in the so-called degenerate cases.

We now present some applications of Theorem 4.1.

Application 4.1 We begin with Example 3.1 in which we considered the periodic boundary value problem

$$\begin{cases} x'' + x & = \lambda(x^3 + f(t)) \\ x(0) - x(2\pi) & = 0 \\ x'(0) - x'(2\pi) & = 0 \end{cases}$$

where f is a continuous function, periodic of period 2π and such that $a_1^2 + b_1^2 > 0$, with a_1 and b_1 the coefficients of Cost and Sin t in the Fourier expansion of f. We have already seen that only the point $(\bar{a}\,\text{Cos}\,t + \bar{b}\,\text{Sin}\,t, 0)$ with

$$\bar{a} = \frac{-\frac{4}{3}a_1}{\sqrt[3]{a_1^2 + b_1^2}}, \qquad \bar{b} = \frac{-\frac{4}{3}b_1}{\sqrt[3]{a_1^2 + b_1^2}}$$

satisfies the necessary condition. It can be easily seen that $\det(V_x(x_0)) > 0$. Hence the connected branch C_+ of non trivial solutions whose closure contains $(\bar{a}\,\text{Cos}\,t + \bar{b}\,\text{Sin}\,t, 0)$ is unbounded.

Without further analysis we are not in a position of saying how the branch becomes unbounded. However, a change of variable suggests that the branch becomes unbounded in the E direction. In fact set $y = x + c$ where c is the constant function $1/\sqrt{\lambda}$. Then y satisfies the differential equations

$$y''(t) - 2y(t) = \sqrt{\lambda}\,(3y^2(t) + \sqrt{\lambda}y^3(t) + \sqrt{\lambda}f(t)).$$

The only periodic solution of this differential equation for $\lambda = 0$ is $\mathbf{0}$ and a branch of non trivial solutions starts from $(\mathbf{0}, 0)$. Hence, our original problem has a branch of solutions coming from infinity as λ approaches 0.

Therefore, using this approach, we cannot say if, for example, the problem has a solution for $\lambda = 1$. Therefore, if we are interested in establishing the existence of a solution of the problem

$$(4.9)\quad \begin{cases} x'' + x & = & x^3 + f(t) \\ x(0) - x(2\pi) & = & 0 \\ x'(0) - x'(2\pi) & = & 0 \end{cases}$$

then a different approach, as proposed in the next application, may be more suitable.

Application 4.2 Following Example 3.2 we consider the one-parameter family of problems

$$\begin{cases} x'' & = & \lambda(-x + x^3 + f(t)) \\ x(0) - x(2\pi) & = & 0 \\ x'(0) - x'(2\pi) & = & 0 \end{cases}$$

As we mentioned previously, there are either 3, 2, or 1 starting points, according to the different values of $\bar{f} = \frac{1}{2\pi}\int_0^{2\pi} f(t)dt$. The sufficient condition is satisfied by all points in the first and last cases. When only two possible starting points are found, one does and the other does not satisfy the sufficient condition. Therefore, we always have un unbounded branch of solutions. Moreover, it can be shown that the branch must intersect $E \times \{1\}$.

In fact, let t_0 be the point such that $\|x\| = |x(t_0)|$, for the pair (x, λ). If t_0 is a maximum point for $x(t)$, then $x(t_0) > 0$ and we obtain $x(t_0) \geq x^3(t_0) + f(t_0)$, while if t_0 is a minimum point for $x(t)$, then $x(t_0) < 0$ and we obtain $x(t_0) \leq x^3(t_0) + f(t_0)$. In either case, we see that we can select a constant M such that $\|x\| \leq M$. Therefore the branch becomes unbounded in the λ direction and it intersects $E \times \{1\}$.

5. Degeneracy and Continuation in Topological Vector Spaces.

Degeneracy

Our first problem, in this part, will be to analyze a case which can be called degenerate. By this we mean that the necessary condition is satisfied by all points of the differentiable manifold M. In the two special cases we have presented in more details in the two previous sections the situation of degeneracy can be described as follows.

In the case when the bifurcation is expected to be from the kernel of the linear operator L, we consider nonlinear maps $H : E \times \mathbf{R} \to F$ such that

$$\frac{\partial QH}{\partial \lambda}(x, 0) = \mathbf{0} \tag{5.1}$$

for all $x \in \text{KerL}$. Our goal is to find pairs of the form $(x, 0)$, $x \in \text{KerL}$, which are starting points of non trivial branches of solutions.

In the case when the bifurcation is expected to be from the line of trivial solutions, we consider linear operators L (see equation 3.1) for which $\dim(\ker L) > 0$, and again look for vectors $(\mathbf{0}, \lambda)$ which are starting points of branches of non trivial solutions.

The following two examples illustrate the kind of problems we have in mind.

Example 5.1 Consider the periodic boundary value problem

$$\begin{cases} x'' + x & = \lambda(x^2 + f(t)) \\ x(0) - x(2\pi) & = 0 \\ x'(0) - x'(2\pi) & = 0 \end{cases} \tag{5.2}$$

where f is such that

$$\int_0^{2\pi} f(t)\text{Cos}t\,dt = \int_0^{2\pi} f(t)\text{Sin}t\,dt = 0$$

Following the analysis of Example 3.1 we can easily verify that $\frac{\partial QH}{\partial \lambda}(x, 0) = \mathbf{0}$ for all $x \in \text{KerL}$, i.e. all $x(t) = a\text{Cos}t + b\text{Sin}t$, $a, b \in \mathbf{R}$. Thus the necessary condition is satisfied by all elements of $\text{KerL} \times \{0\}$, and the theory presented in Sections 3 and 4 cannot be used to find bifurcation points from $\text{KerL} \times \{0\}$.

Example 5.2 Consider the boundary value problem

$$(5.3)\qquad \begin{cases} x'' = \lambda x^3 + x^2 \\ x(0)-x(2\pi) = 0 \\ x'(0)-x'(2\pi) = 0 \end{cases}$$

Then $\mathrm{KerL} = \{\mathbf{x} \in E\colon\ x(t) = c \text{ for all } t \in [0, 2\pi]\}$ and each point of the parameter line $\{(\mathbf{0}, \lambda), \lambda \in \mathbf{R}\}$ is a solution of (5.3). The linear operator L is never an isomorphism, and the necessary condition is satisfied by all pairs $\{(\mathbf{0}, \lambda), \lambda \in \mathbf{R}\}$.

We shall present here how the situation illustrated by Example 5.1 can be analyzed. For a more complete treatment of this topic the interested reader should consult [7].

Let us consider the equation

$$(5.4)\qquad Lx = H(\mathbf{x}, \lambda)$$

where, as in (3.1), L is a Fredholm operator of index zero, acting between the Banach spaces E and F, and $H : E \times \mathbf{R} \to F$ is compact, of class C^1, and such that $H(\mathbf{x}, \lambda) = \lambda R(\mathbf{x})$. The following result holds (see [7]).

Theorem 5.1 *Let* $L : E \to F$ *be a bounded, linear, Fredholm operator of index* 0. *Assume that* $\dim\mathrm{KerL} \geq 1$. *Let* $P : E \to E$, $Q : F \to F$ *be linear projections such that*

$$\mathrm{ImP} = \mathrm{KerL} \quad and \quad \mathrm{KerQ} = \mathrm{ImL}.$$

Let $K : \mathrm{ImL} \to \mathrm{KerP}$ *be a continuous linear right inverse of* L, i.e. $LK\mathbf{y} = \mathbf{y}$ *for all* $\mathbf{y} \in \mathrm{ImL}$ *and* $R : E \to F$ *be continuous, of class* C^2 *in a neighborhood of* KerL *and such that* $QR(\mathbf{x}) = \mathbf{0}$ *for all* $\mathbf{x} \in \mathrm{KerL}$. *Assume that the following two conditions hold at a point* $\mathbf{x}_0 \in \mathrm{KerL}$

$$(5.5)\qquad QR_{\mathbf{x}}(\mathbf{x}_0)KR(\mathbf{x}_0) = \mathbf{0}, \quad and$$

$$(5.6)\qquad QR_{\mathbf{xx}}(\mathbf{x}_0)(KR(\mathbf{x}_0), .) + QR_{\mathbf{x}}(\mathbf{x}_0)K(I - Q)R_{\mathbf{x}}(\mathbf{x}_0)$$

is an isomorphism between ImP *and* ImQ. *Then* $(\mathbf{x}_0, 0)$ *is a bifurcation point for the equation*

$$(5.7)\qquad Lx = \lambda R(\mathbf{x}).$$

<u>Proof.</u> Using the Lyapunov-Schmidt decomposition we can write (5.7) in the equivalent form

$$(5.8)\qquad \begin{cases} \mathbf{v} = \lambda K(I - Q)R(\mathbf{u} + \mathbf{v}) \\ \mathbf{0} = QR(\mathbf{u} + \mathbf{v}) \end{cases}$$

Since R is of class C^2, we see that for λ sufficiently small and for a fixed $\mathbf{u}$, the first equation of (5.8) has a unique solution $\mathbf{v} = g(\mathbf{u}, \lambda)$, which tends to $\mathbf{0}$ as $\lambda \to 0$. Moreover g is differentiable with respect to λ, with continuous partial derivative, and for $\lambda = 0$ we obtain

$$(5.9) \qquad g_\lambda(\mathbf{u}, 0) = K(I - Q)R(\mathbf{u}) = KR(\mathbf{u})$$

since $QR(\mathbf{x}) = \mathbf{0}$ for all $\mathbf{x} \in \mathrm{KerL}$. Substituting $\mathbf{v} = g(\mathbf{u}, \lambda)$ into the second equation of (5.8), we see that, if $(\mathbf{x}_0, 0)$ is a bifurcation point, then, for $\lambda \neq 0$, the $\mathbf{u}$ part must satisfy the equality

$$(5.10) \qquad QR(\mathbf{u} + g(\mathbf{u}, \lambda)) = \mathbf{0}$$

with $\mathbf{u} \to \mathbf{x}_0$ as $\lambda \to 0$. Therefore

$$(5.11) \qquad \lim_{\lambda \to 0} \frac{QR(\mathbf{u}+g(\mathbf{u}, \lambda)) - QR(\mathbf{x}_0)}{\lambda} = QR_x(\mathbf{x}_0)g_\lambda(\mathbf{x}_0, 0) = \mathbf{0}.$$

Substituting (5.9) into (5.11) we obtain (5.5) which is therefore a necessary condition for $(\mathbf{x}_0, 0)$ to be a bifurcation point for (5.7).

At this point, notice that

$$QR_{xx}(\mathbf{x}_0)(KR(\mathbf{x}_0), \cdot\,) + QR_x(\mathbf{x}_0)K(I - Q)R_x(\mathbf{x}_0)$$

is an isomorphism between ImP and ImQ. Therefore, using the Implicit Function Theorem, we can solve (5.10) to find $\mathbf{u} = h(\lambda)$ for λ sufficiently small, and the bifurcating branch is given by

$$(5.12) \qquad \mathbf{u} = h(\lambda), \mathbf{v} = g(\mathbf{u}, \lambda).$$

Notice that in Theorem 5.1 we do not assume that R is compact. Consequently, the obtained conclusion is only a *local* description of the bifurcating branch. No information is provided on its global behavior. The assumption $H(\mathbf{x}, \lambda) = \lambda R(\mathbf{x})$ makes the computations simpler, but it is not necessary.
Therefore the theorem can be extended to the more general case $L\mathbf{x} = H(\mathbf{x}, \lambda)$.

We now want to include the compactness of the non linear part R and study the global behavior of the bifurcating branch.

Theorem 5.2 *Assume that* L *and* R *satisfy the assumptions of* Theorem 5.1 *and that, in addition,* R *is compact. Moreover, suppose that at the point* $(\mathbf{x}_0, 0)$, $\mathbf{x}_0 \in \mathrm{KerL}$, *the two conditions* (5.5) *and* (5.6) *are satisfied. Then there are two connected branches*

$$C_- \text{ and } C_+ \text{ of } \Sigma = \{(\mathbf{x}, \lambda) : L\mathbf{x} = \lambda R(\mathbf{x}), \lambda \neq 0\},$$

such that

i) *the closure of each branch contains the point* $(\mathbf{x}_0, 0)$,
ii) *each branch is either unbounded or its closure contains another point* $(\mathbf{y}_0, 0)$, $\mathbf{y}_0 \in \mathrm{KerL}$, $\mathbf{y}_0 \neq \mathbf{x}_0$.

Proof. We shall give the proof only in the case $\lambda>0$. Only minor changes are needed to adapt it to the case $\lambda<0$.
Define $G(z, \lambda) = z - N(z, \lambda)$ where

$$N(z, \lambda) = \begin{cases} Pz+K(I-Q)R(Pz+\lambda(I-P)z)+\frac{1}{\lambda}JQR(Pz+\lambda(I-P)z) & \text{for } \lambda\neq 0 \\ Pz+K(I-Q)R(Pz)+JQR_z(Pz)(I-P)z & \text{for } \lambda=0 \end{cases}$$

where $R_z(Pz)$ is the map R_z evaluated at the point Pz.
It is not hard to verify, using the hypotheses (5.5) and (5.6), that the operator G satisfies the assumptions of Corollary 4.2 with respect to the pair $(x_0 + KR(x_0), 0)$. Hence there is a connected branch D_+ of non trivial solutions of the equation $G(z, \lambda) = 0$, such that the point $(x_0 + KR(x_0), 0)$ belongs to the closure of D_+ and either one of the two conditions

- D_+ is unbounded or
- its closure contains another point $(z_0, 0)$, such that $G(z_0, 0) = 0$

is satisfied. We define

$$C_+ = \{(x, \lambda) : x = Pz + \lambda(I - P)z \text{ for every } (z, \lambda) \in D_+\}.$$

From $G(z, \lambda) = 0$ with $\lambda > 0$, we obtain

$$\begin{cases} v = \lambda K(I-Q)R(x) \\ 0 = QR(x) \end{cases}$$

Hence the point $x = (Pz + \lambda(I - P)z, \lambda)$ is a solution of $Lx = \lambda R(x)$.

For every point $(z, \lambda) \in D_+$ we have $z = Px + K(I - Q)R(x)$. Therefore, if C_+ is bounded, so is D_+. This shows that the unboundedness of D_+ implies the unboundedness of C_+. Moreover, if a point $(z_0, 0) \neq (x_0, 0)$ belongs to the closure of D_+, then the equality

$$z_0 = Pz_0 + KR(Pz_0) + JQR_z(z_0)((I - P)z_0)$$

implies $z_0 = Pz_0 + KR(Pz_0)$. Hence $(y_0, 0)$, $y_0 = Pz_0$ belongs to the closure of C_+ and $y_0 \neq x_0$. Therefore C_+ satisfies conditions i) and ii) of the theorem.

Notice that in Theorem 5.2 the assumption $H(x, \lambda) = \lambda R(x)$ is not necessary. The proof, with some suitable changes, can be adapted to the case where the the function $H(x, \lambda)$ is not homogeneous of degree 1 with respect to λ.

We now apply Theorem 5.2 to the study of the boundary value problem proposed in Example 5.2. We have already seen that the problem

$$\begin{cases} x'' + x & = \lambda(x^2 + f(t)) \\ x(0) - x(2\pi) & = 0 \\ x'(0) - x'(2\pi) & = 0 \end{cases}$$

can be written as an operator equation of the form (5.4) with $H(x, \lambda) = \lambda R(x)$ and that the degeneracy condition is satisfied provided that $a_1 = b_1 = 0$, where a_1 and b_1 are the coefficients of $\operatorname{Cos} t$ and $\operatorname{Sin} t$ in the Fourier expansion of f. The search for those vectors **u** of KerL, such that the pairs (**u**, 0) satisfy the necessary condition, gives us the system

$$(5.13) \qquad \begin{cases} 5a^3 + 5ab^2 + 12aa_0 - 2bb_2 & = 0 \\ 5a^2b + 5b^3 + 12\,ba_0 - 2ab_2 & = 0 \end{cases}$$

where

$$a_0 = \frac{1}{2\pi}\int_0^{2\pi} f(t)dt \quad \text{and} \quad b_2 = \frac{1}{2\pi}\int_0^{2\pi} f(t)\operatorname{Sin}2t\, dt.$$

The time scale has been adjusted so that $a_2 = \frac{1}{2\pi}\int_0^{2\pi} f(t)\operatorname{Cos}2t\, dt = 0$. Solving system (5.9) we obtain

$$a = b = 0$$

$$(5.14) \qquad a = b = \pm\sqrt{\frac{b_2 - 6a_0}{5}} \quad \text{if } b_2 - 6a_0 \geq 0$$

$$a = b = \pm\sqrt{-\frac{b_2 + 6a_0}{5}} \quad \text{if } b_2 + 6a_0 \leq 0$$

Hence the system has

1 solution if $6a_0 \geq |b_2|$
3 solutions if $6|a_0| \leq |b_2|$ and $6a_0 \neq |b_2|$
5 solutions if $6\, a_0 < -|b_2|$

To apply Theorem 5.2 to our problem we need to verify that the sufficient condition is satisfied. This amounts to check that the Jacobian of (5.13) is different from zero. For the solution $a = b = 0$ we find that this is true if $6|a_0| \neq |b_2|$. For the remaining solutions we must add the condition $b_2 \neq 0$. When these conditions are verified by the Fourier coefficients of the function f, all solutions listed above provide pairs of the form $(x, 0)$, $x \in$ KerL, which are bifurcating points. A further analysis of the sign of the determinant shows that there is always an unbounded branch.

The problem admits a bifurcation from infinity. To see this let us make the substitution

$$(5.15) \qquad x(t) = \frac{1}{\lambda} + y(t)$$

which changes our differential equation into

$$(5.16) \qquad y''(t) - y(t) = \lambda(y^2(t) + f(t))$$

It is easily seen that (5.16) has periodic solutions of period 2π for λ sufficiently small, and the solutions approach $\mathbf{0}$ as $\lambda \to 0$. Consequently (5.2) has a branch of solutions going to ∞ as $\lambda \to 0$.

The analysis presented in this section for the search of bifurcation points among the pairs (x, λ), $x \in \mathrm{KerL}$ in the degenerate case $\frac{\partial QH}{\partial \lambda}(x, 0) = \mathbf{0}$ for all $x \in \mathrm{KerL}$, has not been extended to the case when the bifurcation is expected to be from the line of trivial solutions $(\mathbf{0}, \lambda)$.

Presumably, the two situations are again similar and the two results are a particular case of a more general theory regarding the bifurcation from a differentiable manifold M when all points of M satisfy the necessary condition.

Continuation in locally convex topological vector spaces

We now present an analysis of continuation in locally convex topological vector spaces. Degree theory has been extended to the context of locally convex spaces, but one would be mistaken in assuming that the continuation principle could be established by simply replacing a Banach space E with an arbitrary locally convex (or even Fréchet) space. For a more complete discussion of this topic the reader is referred to [16, 17], from which the ideas of this part are derived.

The homotopy property of the degree introduced in the third section, can be rephrased in the following form

Theorem 5.3 (Leray-Schauder Continuation Principle) *Let* U *be an open subset of a Banach space* E, *and* $H : \bar{U} \times [0, 1] \to E$ *be a continuous and compact map* (*i.e. sending bounded sets into relatively compact sets*). *Assume that*

i) $H(x, 0) = x_0 \in U$ *for all* $x \in \bar{U}$;

ii) $\Sigma = \{x \in \bar{U}: x = H(x, \lambda)$ *for some* $\lambda \in [0, 1]\}$ *is bounded and* $\Sigma \cap \partial U = \varnothing$.

Then the equation $x = H(x, 1)$ *has a solution in* U.

The above principle is based on the definition of degree for a continuous and compact vector field $T = I - f : \bar{U} \to E$, such that the set of fixed point of f is bounded and does not intersect the boundary of U. When we examine the extensions of degree theory to locally convex spaces we notice that the different authors introduce some additional assumptions, which, at a first glance, may seem superfluous. For example, Nagumo, M. [33] assumes that $\overline{f(U)}$ is compact, while Schwartz, J. T. [37] assumes that U is finitely bounded in the sense that its intersection with every finite dimensional subspace of E is bounded. These assumptions have the side effect of limiting considerably the applicability of the results. In the case of Nagumo one finds very few maps for which the theory can be used; in the case of Schwartz, one can hardly find open sets with the required property.

Therefore, we may ask why this additional and significantly restrictive assumptions are introduced. After all, can' t we proceed just as in the case of Banach spaces ? The following simple example shows that the answer to this question is negative, and something else need to be done to make the things work.

Denote by F the Fréchet space $C[0, \infty)$, whose elements are the continuous real functions on the half line. F is given the standard topology of uniform convergence on compact sets. Define the following map $H : F \times [0, 1] \to F$

$$H(x, \lambda)(t) = \lambda \int_0^t (1 + x^2(s))ds \qquad (5.17)$$

The map H is continuous and it sends bounded sets of $F \times [0,1]$ into relatively compact subsets of F. For $\lambda = 0$, the only solution of $H(x, 0) = \mathbf{0}$ is $x = \mathbf{0}$, and one can easily verify that (5.17) has no solutions for $\lambda > 0$. Thus, by selecting as U any open neighborhood of the origin we see that the two conditions i) and ii) of Theorem 5.3 are satisfied by H. However the conclusion of the theorem fails.

This example tells us a lot of things. First, there are no finitely bounded open sets in F. Hence, in this important space, the extension of Schwartz is useless! Second, this map is not compact in the sense of Nagumo, and it is hard to imagine what kind of maps, besides the constants, could satisfy his assumption.

In trying to understand what is *lost* in passing from Banach spaces to Fréchet spaces, or more generally to locally convex topological vector spaces, we may observe that in a Banach space the assumption "Σ *is bounded and* $\partial U \cap \Sigma = \varnothing$" allows us to confine our attention to bounded open sets. Therefore, a continuous and compact (i.e. which sends bounded sets into relatively compact sets) map $f : \bar{U} \to E$ can be considered as having the property that the closure of its entire image is compact. This situation changes in Fréchet spaces, simply because a Hausdorff locally convex space does not admit bounded open sets, unless the space is normable. Thus, something different need to be done in this class of spaces, since the two approaches presented above do not seem to be of much help.

The result we are going to present is contained in [16], where more general continuation principles are proved and applications are given to boundary value problems in non-compact intervals. The continuation principle is new also in the case where the space is Banach, although no applications are provided in this context.

Theorem 5.4 *Let* E *be a Hausdorff locally convex topological vector space,* Q *be a convex closed subset of* E *and* $T : Q \times [0,1] \to E$ *be a continuous map with a relatively compact image. Assume that*

j) $T(., 0) : Q \to Q$;

jj) *if* $(x, \lambda) \in \partial Q \times [0,1]$ *and* $T(x, \lambda) = x$, *then there exists an open neighborhood* U_x *of* x *in* E *and* I_λ *of* λ *in* $[0, 1]$ *such that* $T(y, \mu) \in Q$ *for all* $(y, \mu) \in (U_x \cap \partial Q) \times [0, 1]$.

Then the equation $T(x, 1) = x$ *has a solution in* Q.

Proof. We shall establish the result in the case when Q is a convex and closed subset of a Banach space E. The interested reader can find the proof of the general case in the paper of Furi, M. - Pera, M. P. [16].

Denote by $r : E \to Q$ a continuous retraction and let $N(x, \lambda) = T(r(x), \lambda)$. Since T has a relatively compact image, we can assume, without loss of generality, that Q is bounded and we can find a convex, closed and bounded set Q_0, with non empty interior, such that

$$Q \subset Q_0,\ N(., \lambda) : Q_0 \to Q_0, \text{ for all } \lambda \in [0, 1] \text{ and } N(x, \lambda) \neq x \text{ for all } x \in \partial Q_0.$$

Hence for every $\lambda \in [0, 1]$ the set of fixed points of $N(., \lambda)$, denoted by Σ_λ, is not empty . We need to show that $\Sigma_1 \cap Q \neq \emptyset$.

By contradiction assume that

$$\Sigma_1 \cap Q = \emptyset \text{ and let } d = \inf\{ \|x - y\| : x \in Q, y \in \Sigma_1\}.$$

Then $d > 0$ and for every n such that $1 < dn$, let $V_n = \{z \in E : d(z, Q) < n^{-1}\}$. The set V_n is open, bounded and convex. By Theorem 2.1 there exists a connected component of the set

$$\Sigma = \{(x, \lambda) \in K \times [0, 1] : N(x, \lambda) = x\}$$

which joins Σ_0 with Σ_1. Hence we can find $(x_n, \lambda_n) \in \partial V_n \cap \Sigma$.

As $n \to \infty$, $\lambda_n \to \lambda_*$ and $x_n \to x_* \in Q$. Hence

$$r(x_*) = x_* \, , \lambda_* < 1 \text{ and } T(x_*, \lambda_*) = x_*.$$

Since $r(x_n) \in \partial Q$ for every n and $r(x_n) \to x_*$ as $n \to \infty$, we obtain, from assumption jj) that $T(r(x_n), \lambda_n) \in Q$ for all but finitely many n's. Combining this result with the equalities

$$N(x_n, \lambda_n) = x_n$$

we get $r(x_n) = x_n$. Since $x_n \in \partial V_n$ for every n, we have reached a contradiction.

Notice that the above continuation principle lacks stability with respect to small perturbations, in the sense that maps arbitrarily close to T may fail to satisfy the required assumption and consequently, violate the conclusion of the theorem. The following example illustrates this situation.

Example 5.3 Let $T : \mathbf{R}^2 \times \mathbf{R} \to \mathbf{R}^2$ be defined by $T(x, y, \lambda) = (\lambda x, 0.5\lambda)$. Assume that $Q=[-1,1] \times [-0.5, 0.5]$. Then $T(Q \times [0,1]) \subset Q$. Therefore T satisfies the assumptions of Theorem 5.4.

Let $\varepsilon > 0$ and define $T_\varepsilon(x, y, \lambda) = (\lambda x, 0.5\lambda + \varepsilon)$. For ε sufficiently small and for $\lambda = 0$ we have $T_\varepsilon(x, 0) \in Q$. However, the equation $T_\varepsilon(x, 1) = x$ does not have any solution in Q, since the required value for y would be $0.5 + \varepsilon$, while the maximum value of y in Q does not exceed 0.5.

6. A Priori Bounds

In this last section we want to analyze recent results established by Capietto, A. - Mawhin, J.- Zanolin, F. [2, 3] and Furi, M. - Pera, M. P. [18, 19, 20] in which a priori bounds are incorporated into continuation principles in order to insure the existence of solutions with specific properties. The approaches of the two groups of authors have similarities and differences. Our aim here is to provide the reader with the guiding ideas used by them, together with a continuation principle which attempts to combine the advantages of both approaches.

Our standing assumptions are the following.

- L is a Fredholm operator of index 0 acting between two Banach spaces E and F.
- $N : E \times \mathbf{R} \to F$ is nonlinear, continuous and compact.
- The solution set Σ_0 of the equation $Lx = N(x, 0)$ is not empty.

We are interested in sufficient conditions which will insure the existence of a connected branch S of solutions of the equation

$$(6.1) \qquad Lx = N(x, \lambda)$$

with $\lambda \neq 0$, such that

(6.2) a) $\bar{S} \cap \Sigma_0 \neq \varnothing$, b) $\mathbf{0} \times [0, d] \subset \pi_2(\bar{S})$

for some $d > 0$ where $\pi_2(x, \lambda) = (\mathbf{0}, \lambda)$. By solution we mean a pair (x, λ) which satisfies equation (6.1). In the case when $\lambda \neq 0$ the pair is called non-trivial.

A sufficient condition for (6.2) is usually given in two steps .

- The first guarantees the existence of a connected branch S which satisfies the inequality (6.2 a) and has the property $\mathbf{0} \times [0, \delta) \subset \pi_2(\bar{S})$ for some $\delta \in (0, d)$.
- The second insures (6.2 b).

The solution set Σ_0 could be bounded or unbounded. Here we shall assume this second alternative. However, we also assume that problem (6.1) can be rewritten in such a way that the set of its solution pairs remains the same for $\lambda > 0$, while for $\lambda = 0$ we can find an open bounded set U such that the new solution set, which we shall indicate with S_0 has the property

(6.3) $S_0 \cap U \neq \varnothing, \quad S_0 \cap \partial U = \varnothing$.

In the case when $N(x, \lambda) = \lambda M(x)$, the modification to problem (6.1) could take the form

(6.4) $Lx + QM(x) = \lambda(I - Q)M(x)$

where $Q : F \to F$ is a linear projection such that $\operatorname{Ker} Q = \operatorname{Im} L$. The solution pairs of problem (6.4) and (6.1) coincide for $\lambda > 0$, but they are different for $\lambda = 0$, since

(6.5) $\Sigma_0 = \operatorname{Ker} L \times \{0\}, \quad S_0 = \{(x, 0) \in \operatorname{Ker} L \times \{0\} : M(x) \in \operatorname{Im} L\}$.

We denote the modified problem by

(6.7) $Lx + T(x) = H(x, \lambda)$

where the map T may be nonlinear. We rewrite (6.7) as

(6.8) $Lx = R(x, \lambda)$

Let $P : E \to E$ be a projection such that $\operatorname{Im} P = \operatorname{Ker} L$ and $K : \operatorname{Im} L \to \operatorname{Ker} P$ be linear, bounded and such that $LKy = y$ for all $y \in \operatorname{Im} L$. Then (6.8) becomes equivalent to

(6.9) $x = Px + JQR(x, \lambda) + K(I - Q)R(x, \lambda) = V(x, \lambda)$

where $J : \mathrm{Im}\, Q \to \mathrm{Ker}\, P$ is a linear isomorphism. Notice that V is continuous and compact.

A sufficient conditions for starting the connected branch can now be given in terms of the degree of the map $I - V(.,0)$ on the bounded set U, by requiring that

$$(6.10) \qquad \deg(I - V(.,0), U, \mathbf{0}) \neq 0.$$

Then, using the same reasoning as in the proof of Theorem 4.1 we can show that there is a connected branch $S \subset E \times [0,\infty)$ of non-trivial solutions such that $\overline{S} \cap S_0 \cap U \neq \emptyset$ and

- either S is unbounded;
- or $\overline{S} \cap S_0 \cap U^c \neq \emptyset$.

Our goal is to insure that the first alternative holds and $\mathbf{0} \times [0, d] \subset \pi_2(\overline{S})$. The simplest way to avoid the second alternative would be to require that there exists a countable family of disjoint, open and bounded sets $\{U_n : n = 1,..\}$ such that

$$(S_0 \cap U^c) \subset \cup\, U_n$$

and

$$(6.11) \qquad \deg(I - V(.,0), U_n, \mathbf{0}) \deg(I - V(.,0), U, \mathbf{0}) \geq 0.$$

In fact, denote by

$$(6.12) \qquad V_n = \bigcup_{i=1}^{n} U_i$$

Then $\deg(I - V(.,0), V_n, \mathbf{0}) \deg(I - V(.,0), U, \mathbf{0}) \geq 0$. This inequality shows that $\overline{S} \cap V_n = \emptyset$.

To obtain the inclusion $\mathbf{0} \times [0, d] \subset \pi_2 \overline{S}\,)$ we need some kind of control on the growth of the connected branch S.

To achieve this goal let $g : E \times [0,\infty) \to \mathbf{R}$ be continuous and such that for every $(\mathbf{x}, \lambda)$ solution of (6.9) with $\lambda < d$ one has

$$(6.13) \qquad |\, g(\mathbf{x}, \lambda)| \geq h(\|\mathbf{x}\|)$$

where $h : \mathbf{R} \to \mathbf{R}$ has the property $h(r) \to \infty$ as $r \to \infty$. Combining (6.13) with some other suitable assumptions on g can guarantee the result we want. For this purpose denote by Σ the set of solutions of (6.9) and assume that there exists a constant $M > 0$ such that $g(\mathbf{x}, \lambda)$ is a locally constant function on Σ whenever $\|\,\mathbf{x}\,\| \geq M$.

Let us suppose that the set $C_d = \{x \in E : (x, \lambda) \in S$ for some $\lambda < d\}$ is unbounded. By (6.13) the restriction of g to those elements (x, λ) of $\bar{S}$ such that $\lambda < d$ must contain a half line of the form $[a, +\infty)$. This however is impossible since g is locally constant on Σ for $\| x \| \geq M$ and $\bar{S}$ is connected. Consequently, C_d must be bounded. Therefore, by considering only those elements of $\bar{S}$ such that $\lambda < d$ we obtain that both projections of $\bar{S}$ on E and on **R** are bounded. Since $\bar{S}$ is unbounded it must have elements (x, λ) with $\lambda \geq d$.

Consequently $\mathbf{0} \times [0, d] \subset \pi_2(\bar{S})$ and the equation (6.1) has a solution for $\lambda = d$.

Summarizing the situation presented so far we can state the following theorem

Theorem 6.1 *Let* E, F *be Banach spaces,* $N : E \times \mathbf{R} \to F$ *be continuous and compact, in the sense that the image of any bounded set of* $E \times \mathbf{R}$ *is totally bounded in* F, *and let* $L : E \to F$ *be a bounded Fredholm operator of index* 0. *Denote by* Σ *the set*

$$\Sigma = \{(x, \lambda) : Lx = N(x, \lambda)\} \text{ and let } \Sigma_\lambda = \Sigma \cap (E \times \{\lambda\}).$$

Assume that the equation $Lx = N(x, \lambda)$ *can be replaced by* $Lx = R(x, \lambda)$ *with* R *continuous and compact, so that* Σ_λ *remains unchanged for* $\lambda > 0$, *while, for* $\lambda = 0$ *the new solution set* $S_0 \subset \Sigma_0$ *has the following properties.*

- *There exists an open, bounded set* U *of* E *such that* $P_1(S_0) \cap U \neq \emptyset$, *where* $P_1(x, \lambda) = x$.
- $P_1(S_0) \cap \partial U = \emptyset$.
- $\deg(L - R(., 0), U, \mathbf{0}) \neq 0$.
- *There is a countable family of disjoint, open bounded sets* U_n *such that* $(S_0 \cap U^c) \subset \cup U_n$ *and* $\deg(L - R(., 0), U_n, \mathbf{0}) \deg(L - R(., 0), U, \mathbf{0}) \geq 0$.

Denote by Σ^* *the solution set of the equation* $Lx = R(x, \lambda)$ *and suppose that there exists a continuous function* $g : E \times \mathbf{R} \to E$ *such that*

- $|g(x, \lambda)| \geq h(\|x\|)$ *for every* $(x, \lambda) \in \Sigma^*$ *with* $\lambda < d$ *and* $h(r) \to \infty$ *as* $r \to \infty$.
- *There exists a positive real number* M *such that* g *is a locally constant function on* Σ^* *for* $\| x \| \geq M$.

Then the equation $Lx = N(x, d)$ *has a solution.*

The degree $\deg(L - R(.,0), U, \mathbf{0})$ is the coincidence degree developed by Mawhin, J. [31, 21]. It is defined (see (6.10)) to be equal to $\deg(I - V(., 0), U, \mathbf{0})$ and it is independent of the projections P, Q and of the isomorphism J. For applications of the previous result to the existence of periodic solutions of differential equations the reader should consult the papers of Capietto, A.- Mawhin, J. - Zanolin, F. [2, 3] and Furi, M. - Pera, M. P. [18, 19, 20].

7. Some open questions

I have mentioned some open questions in the course of these lectures, like the incorporation of bifurcation and cobifurcation problems into a unified theory.

I have presented an approach to degeneracy in the cobifurcation setting. I am not aware that anything similar has been done so far for the bifurcation setting. A typical situation would probably be the study of branches of non-trivial solutions of nonlinear operator equations of the form $L(\mathbf{x}, \lambda) = R(\mathbf{x}, \lambda)$, where $R(\mathbf{0}, \lambda) = \mathbf{0}$ for every λ and L is never an isomorphism.

In the bifurcation setting there is, so to speak, a natural way to explore the existence of branches coming from infinity, with a transformation which reduces the problem to the existence of branches bifurcating from the line $(\mathbf{0}, \lambda)$. Can anything similar be done for the cobifurcation setting? I have presented, in this notes, two particular cases in which a suitable change of variable can detect this feature. The two changes are different. Can a general theory be provided ?

References

1. Alexander, J.C. (1981): *A primer on connectivity.* In Proceedings of the Conference in Fixed Point Theory, Fadell, E. - Fournier, G. Editors, Springer-Verlag Lecture Notes in Mathematics, Vol. **886**, 455-488.

2. Capietto, A. - Mawhin, J. - Zanolin, F. (1989): *A continuation approach to superlinear periodic boundary value problems*. Journal of Differential Equations, **88**, 347-395.

3. Capietto, A. - Mawhin, J. - Zanolin, F. (1991): *The coincidence degree of some functional differential operators in spaces of periodic functions and related continuation theorems.* In Delay Differential Equations and Dynamical Systems, Busenberg, S. - Martelli, M. (Editors), Springer-Verlag Lecture Notes in Mathematics, **1475**, 76-88.

4. Darbo, G. (1955) : *Punti uniti in trasformazioni a codominio non compatto.* Rendiconti del Seminario Matematico dell'Università di Padova, **24**, 353-367.

5. Dugundji, J. (1951) : *An extension of Tietze's theorem.* Pacific Journal of Mathematics, **1**, 353-367.

6. Eilenberg, S. - Montgomery, D. (1946): *Fixed point theorems for multivalued transformations.* American Journal of Mathematics, **68**, 214-222.

7. Fabry, C.- Martelli, M. (1986) : *Cobifurcation dégénérée et problèmes aux limites.* Annales de la Societé Scientifique de Bruxelles, **100**, III, 105-114.

8. Fitzpatrick, P. M. - Pejsachowicz, J. (1988) : *Fundamental group of the space of Fredholm operators and global analysis of semilinear equations.* Contemporary Mathematics, **72**, 47-88.

9. Fitzpatrick, P. M. - Pejsachowicz, J. (1991) : *Nonorientability of the index bundle and several parameters bifurcation.* To appear in Journal of Functional Analysis.

10. Fournier, G. - Martelli, M. (1990) : *Boundary conditions and vanishing index for α-contractions and condensing maps.* In Nonlinear Functional Analysis, Milojevic, P.S. (Editor), Lecture Notes in Pure and Applied Mathematics, Marcel Dekker Inc., **21**, 31-49.

11. Fournier, G. - Violette, D. (1986) : *A fixed point index for composition of acyclic maps in Banach spaces.* In Operator Equations and Fixed Point Theorems, Singh, S.P. - Sehgal, V.M. - Burry, J.H.W. (Editors), The MSRI Korea Publications, **1**, 139-158.

12. Furi, M. - Martelli, M. - Vignoli, A. (1980) : *On the solvability of operator equations in normed spaces.* Annali di Matematica Pura e Applicata, **124**, 321-343.

13. Furi, M. - Pera, M. P. (1979) : *On the existence of an unbounded connected set of solutions for nonlinear equations in Banach spaces.* Atti dell'Accademia Nazionale dei Lincei, Rendiconti della Classe di Scienze Matematiche, Fisiche e Naturali, **47**, 31-38.

14. Furi, M. - Pera, M. P. (1983) : *Cobifurcating branches of solutions of nonlinear eigenvalue problems in Banach spaces.* Annali di Matematica Pura e Applicata, **135**, 119-132.

15. Furi, M. - Pera, M. P. (1984) : *Global branches of periodic solutions for forced differential equations on nonzero Euler characteristic manifolds.* Bollettino dell'Unione Matematica Italiana, Analisi Funzionale e Applicazioni, Serie VI, **3**, 157-170.

16. Furi, M. - Pera, M. P. (1987) : *A continuation method on locally convex spaces and applications to ordinary differential equations on noncompact intervals.* Annales Polonici Mathematici, **57**, 331-346.

17. Furi, M. - Pera, M. P. (1987) : *On the fixed point index in locally convex spaces.* Proceedings of the Royal Society of Edinburgh, **106**A, 161-168.

18. Furi, M. - Pera, M. P. (1990) : *On the existence of forced oscillations for the spherical pendulum.* Bollettino dell'Unione Matematica Italiana, **4**-B, 381-390.

19. Furi, M. - Pera, M. P. (1991) : *A continuation principle for the spherical pendulum.* Fixed Point Theory and Applications, Théra, M.A.-Baillon, J.B. (Editors), Pitman Research Notes In Math. 252, Longman Scientific and Technical, 141-154

20. Furi, M. - Pera, M. P. (1991) : *The forced spherical pendulum does have forced oscillations.* In Delay Differential Equations and Dynamical Systems, Busenberg, S. - Martelli, M. (Editors), Springer-Verlag Lecture Notes in Mathematics, **1475**, 176-183.

21. Gaines, R. E. - Mawhin, J. L. (1977) : Coincidence Degree and Nonlinear Differential Equations. Springer-Verlag Lecture Notes in Mathematics, **568**.

22. Granas, A. (1962) : *The theory of compact vector fields and some of its applications to the topology of functional spaces.* Rozpravy Mat., **20**, 1-93.

23. Iannacci, R. - Martelli, M. (1986) : *Branches of solutions of nonlinear operator equations in the atypical bifurcation case.* In Operator Equations and Fixed Point Theorems, Singh, S.P. - Sehgal, V.M. - Burry, J.H.W. (Editors), The MSRI Korea Publications, **1**, 13-24.

24. Kakutani, S. (1941) : *A generalization of Brower fixed point theorem.* Duke Mathematical Journal, **8**, 457-459.

25. Kuratowski, C. (1930) : *Sur les espaces complètes.* Fundamenta Mathematicae, **15**, 292-304.

26. Leray, J. - Schauder, J. (1934) : *Topologie et équations functionnelles.* Ann. Sci. Ec. Norm. Sup., **51**, 45-78.

27. Martelli, M. (1970) : *A lemma on maps of a compact topological space and an application to fixed point theory.* Rendiconti dell'Accademia Nazionale dei Lincei, **49**, 242-243 .

28. Martelli, M. (1981) : *Existence and bifurcation for nonlinear operator equations: an elementary approach.* In Differential Equations and Applications in Ecology, Epidemics and Population Problems, Busenberg, S. - Cooke, K. Editors, Academic Press, 289-304.

29. Martelli, M. (1983) : *Large oscillations of forced nonlinear differential equations.* In AMS Contemporary Mathematics, **21**, 151-159.

30. Martelli, M. (1986) : *Positive eigenvectors of wedge maps.* Annali di Matematica Pura e Applicata, 1**45**, 1-32.

31. Mawhin, J. (1972) : *Equivalence Theorems for nonlinear operator equations and coincidence degree theory for some mappings in locally convex topological vector spaces.* Journal of Differential Equations, **12**, 610-636.

32. Mawhin, J. (1977) : *Functional analysis and boundary value problems.* Studies in Ordinary Differential Equations, Hale, J. (Editor), MAA Studies in Mathematics, **14**, 122-168.

33. Nagumo, M. (1951) : *Degree of mappings in convex, linear topological spaces.* American Journal of Mathematics, **73**, 497-511.

34. Prodi, G. - Ambrosetti, A. (1973) : Analisi non Lineare. Quaderni della Scuola Normale Superiore di Pisa, I Quaderno.

35. Rabinowitz, P. M. (1971) : *Some global results for nonlinear eigenvalue problems.* Journal of Functional Analysis, **7**, 487-513.

36. Sadovskii, B. N. (1972) : *Limit-compact and condensing operators.* Russian Mathematical Surveys, **27**, 85-156.

37. Schwartz, J. T. (1969) : Nonlinear Functional Analysis. Gordon and Breach Science Publishers, New York.

Topological Degree and Boundary Value Problems for Nonlinear Differential Equations

J. Mawhin

Institut mathématique, Université de Louvain, B-1348
Louvain-la-Neuve, Belgium

Dedicated to Jean Leray on the occasion of his 85th birthday

1 Introduction

In a survey paper entitled *Les problèmes non linéaires* [75], LERAY summarizes as follows the new method he has just created, with SCHAUDER [76], to study an important class of nonlinear equations in Banach spaces (our English translation) : 'To be able to assert that the equation $x+\mathcal{F}(x)=0$ is solvable, it is sufficient to prove that it has no *arbitrarily large* solution when one reduces it continuously to an equation like $x=0$. To prove that a functional equation has solutions is hence reduced to the following problem : to obtain a priori bounds for the possible solutions. It would be indeed impossible to imagine that one could solve an equation by a procedure which does not provide some information on the size of the unknowns. For us, to solve an equation consists in bounding the unknowns and precising their shape as much as possible; it is not to construct, through complicated developments, a solution whose practical use will almost always be impossible. One can consider this existence theorem as being a generalization to the nonlinear case of the Fredholm alternative : let $x+\mathcal{L}(x)=b$ (where $\mathcal{L}(x)=\int K(s,s')x(s')\,ds'$ is completely continuous) be a Fredholm equation; this equation has always a solution, except if the equation $x+\mathcal{L}(x)=0$ has some; and this case is just the one where the proposed equation would admit *arbitrarily large* solutions.'

The success of this *Leray-Schauder method* in the study of various types of nonlinear differential, integral or other functional equations has been extraordinary. LERAY himself gave striking applications in fluid mechanics and in elliptic partial differential equations. Many other mathematicians contributed in applying the method to other types of partial differential equations and integral equations. As far as ordinary differential equations (the only applications we shall consider here) are concerned, one should mention an early contribution of DOLPH [31] who uses the Leray-Schauder theorem to prove that the nonlinear

Hammerstein integral equation

$$\psi(x) = \int_a^b K(x,y) f[y, \psi(y)]\, dy,$$

has at least one solution whenever K is the kernel of a completely continuous Hermitian operator in $L_2([a,b])$ with characteristic values

$$\ldots \leq \lambda_{-2} \leq \lambda_{-1} < 0 < \lambda_0 \leq \lambda_1 \lambda_2 \leq \ldots,$$

and the continuous function f satisfies, for some integer N and some reals

$$A,\ y_0 > 0,\ \lambda_N < \mu_N < \mu_{N+1} < \lambda_{N+1},$$

the conditions

$$\mu_N y - A \leq f(x,y) \leq \mu_{N+1} y + A \text{ for } y > y_0,$$
$$\mu_{N+1} y - A \leq f(x,y) \leq \mu_N y + A \text{ for } y < y_0.$$

This result can of course be immediately applied to the two-point boundary value problem

$$\psi''(x) + f(x, \psi(x)) = 0, \psi(a) = \psi(b) = 0,$$

by reducing it to an integral equation of the above type through the Green function. Notice that the conditions upon f just become, when f is linear in y, the Fredholm nonresonance condition mentioned in LERAY's remarks quoted above. One should notice that Dolph's result does not really require the full power of the Leray-Schauder method but can as well be proved using Schauder's fixed point theorem. In 1952, STOPPELLI used the Leray-Schauder theorem to prove the existence of at least one T-periodic solution of the ordinary differential equation

$$y'' + y'|y'| + q(t)y' + y - p^2(t)y^3 = f(t),$$

when q, p and f are continuous and T-periodic and p is positive. He deforms the equation to the simpler one

$$y'' + y'|y'| + y' + y - y^3 = 0,$$

which has only the three isolated T-periodic solutions $y_1 = 0, y_2 = 1, y_3 = -1$, and shows the existence of a priori estimates for the possible T-periodic solutions. The interested reader can find historical and bibliographical information about the subsequent applications of the Leray-Schauder theory to boundary value problems for ordinary differential equations in [63], [122], [124], [48], [64], [41], [89], [90], [91].

The aim of those lectures is to describe some recent developments of the application of Leray-Schauder method to the study of boundary value problems for ordinary differential equations. In Section 2, we shall rapidly describe a formulation of this theory in the setting of some relatively compact perturbations of linear Fredholm operators of index zero, which is well adapted to the applications we have in mind. In this setting, the reduction of the problem to a fixed point problem is done once for all and one can concentrate on the main aspects of the method : find a continuous deformation of the problem into a simpler one for which the corresponding topological degree can be computed and in such a way that a priori estimates hold for the possible solutions of the whole deformation. Those two aspects are of course closely connected : any new result on the computation of the degree allows the use of new deformations for which the a priori estimates can be more easily obtained.

In Section 3, we prove that, under very general conditions, the topological degree associated to the T-periodic solutions of an arbitrary autonomous n-dimensional differential system

$$x' = g(x)$$

is equal to the Brouwer degree in $\mathbf{R}^n$ of the right member g of the equation. This result was first proved in [10] and we present here the proof given in [4] based upon a computation of the Leray-Schauder degree for mappings equivariant with respect to a representation of S^1. Applications of this result to various existence theorems of the perturbational or global type for T-periodic solutions of nonautonomous systems are then given in Section 4.

Section 5 is devoted to the study of some boundary value problems for second order differential equations of the type

$$x'' + g(x) = p(t),$$

with g superlinear, i.e. such that $\frac{g(x)}{x} \to +\infty$ when $|x| \to \infty$, and for corresponding Hamiltonian systems. The important feature of those problems is that the set of possible solutions of the corresponding deformation to the corresponding autonomous equation $x'' + g(x) = 0$ is not a priori bounded. In [12] this difficulty is overcome by introducing a second invariant related to the Poincaré's index of the curve (x, x') and showing an a priori bound for the possible solutions for which this invariant has a fixed value.

Section 6 shows that the Leray-Schauder approach can be applied to the study of periodic solutions of some second order differential equations

$$x'' + cx' + g(x) = p(t),$$

when the restoring force g has a singularity. Both the case of an attractive and of a repulsive force are considered. Using KRASNOSEL'SKII's concept of bifurcation from infinity like in [97] and [98], we add some multiplicity results for the solutions to the recent existence theorems of LAZER-SOLIMINI [74] and HABETS-SANCHEZ ([54],[55]).

Section 7 describes some recent results of ORTEGA ([111],[112]) which shows how topological degree arguments can be used to obtain information about the stability of the periodic solutions of some two-dimensional systems and second order equations, and in particular under AMBROSETTI-PRODI type conditions. We refer to the introductions of the various sections for a more detailed description of their contents.

2 A degree theory for some non linear perturbations of Fredholm operators

2.1 Introduction

Continuation theorems for the study of boundary value problems for ordinary differential equations are often based upon an equivalent formulation as an equation in an abstract space and an application to that equation of some degree theory. The most fundamental setting for a degree theory for mappings in a Banach space is the Leray-Schauder degree for compact perturbations of the identity. On the other hand, the natural abstract formulation of boundary value problems for ordinary differential equations usually leads to an abstract operator which is the sum of a Fredholm linear mapping of index zero and a nonlinear mapping having some properties of relative compactness with respect to the linear mapping. Those problems, as we shall see, can always be reduced to equivalent equations of the Leray-Schauder type but, to avoid this reduction, we shall develope once for all a degree theory for this class of mappings. This was done in [85] using some Liapunov-Schmidt-type techniques. We shall use here a different and somewhat simpler approach. The reader who is familiar with [48] or [89] can of course proceed directly to Section 3.

2.2 Fredholm and L-compact mappings

Let X and Z be real normed vector spaces. A linear mapping $L : D(L) \subset X \to Z$ is called *Fredholm* if the following conditions hold :

(i) $\ker L : L^{-1}(0)$ has finite dimension;

(ii) $R(L) = L(D(L))$ is closed and has finite codimension.

The *index* of L is the integer $\dim \ker L - \operatorname{codim} R(L)$. One can consult [89] for more details on Fredholm operators and their link with linear differential operators with linear boundary conditions.

If L is Fredholm of index zero, there exists a continuous projector $P : X \to X$, a projector $Q : Z \to Z$ such that $R(P) = \ker L$, $\ker Q = R(L)$, and a bijection $J : \ker L \to R(Q)$. It is then easy to verify that $L + JP : D(L) \to Z$ is a bijection. Denote by $\mathcal{F}(L)$ the set of linear continuous mapppings of finite rank $A : X \to Z$ which are such that $L + A : D(L) \to Z$ is a bijection.

Let $\Delta \subset X$ and $N : \Delta \to Z$ a mapping. It is clear that for each $A \in \mathcal{F}(L)$, the equation

$$Lx + Nx = 0 \tag{1}$$

is equivalent to the equation

$$(L + A)x + (N - A)x = 0,$$

and hence to the fixed point problem

$$x + (L + A)^{-1}(N - A)x = 0. \tag{2}$$

Notice that equation (1) is to be considered in $D(L) \cap \Delta$ and is therefore equivalent to (2) in this set; on the other hand, (2) is defined in Δ but as $(L+A)^{-1}(N-A) \in D(L)$, for each $x \in \Delta$, every solution of (2) belongs to $D(L)$ and hence we have the equivalence between (1) considered in $D(L) \cap \Delta$ and (2) considered in Δ. Notice also that $(L+A)^{-1}A$ is always a linear, continuous operator of finite rank in X, and hence a compact operator.

Let E be a metric space and $G : E \to Z$ be a mapping.

Lemma 2.1. *If there exists some $A \in \mathcal{F}(L)$ such that $(L+A)^{-1}G$ is compact on E, then the same is true for any $B \in \mathcal{F}(L)$.*

Proof. Let $B \in \mathcal{F}(L)$; then,

$$(L+B)^{-1}G = (L+B)^{-1}(L+A)(L+A)^{-1}G =$$

$$(L+B)^{-1}(L+B+A-B)(L+A)^{-1}G = [I+(L+B)^{-1}(A-B)](L+A)^{-1}G.$$

As $A-B$ is continuous and has finite rank, $(L+B)^{-1}(A-B)$ is continuous and has finite rank, and hence $(L+B)^{-1}(A-B)(L+A)^{-1}G$ is compact on E. ■

Lemma 2.1 justifies the following definition, introduced in [136], and equivalent to an earlier one in [86].

Definition 2.1. *We say that $G : E \to Z$ is* L-compact *on E if there exist $A \in \mathcal{F}(L)$ such that $(L+A)^{-1}G : E \to X$ is compact on E.*

For $E \subset X, X = Z$ and $L = I$, this concept reduces to the classical one of compact mapping introduced by SCHAUDER [126].

Using the projectors P and Q introduced above and letting

$$K_{PQ} = (L|_{D(L)\cap \ker P})^{-1}(I-Q),$$

(K_{PQ} is the *right inverse of* L associated to P and Q), it is easy to verify that $G : E \to Z$ is L-compact on E if and only if $QG : E \to Z$ is continuous, $QG(E)$ is bounded and $K_{PQ}G : E \to X$ is compact. Of course, if $L : D(L) \to Z$ is invertible, one can take $A = 0$ in Definition 2.1 and the L-compactness of G on E is equivalent to the compactness of $L^{-1}G$ on E.

Remark 2.1. It is clear that if $C \subset E$ and if $G : E \to Z$ is L-compact on E, it is also L-compact on C. Also, if $H : E \to Z$ is L-compact on E, the same is true for $G + H$.

If $G : X \to Z$ is L-compact on each bounded set $B \subset X$, we shall say that G is *L-completely continuous on* X. The following useful property of linear L-completely continuous mappings is proved in [66] using the definition of L-compactness given in [81].

Proposition 2.1. *If $A : X \to Z$ is linear, L-completely continuous on X and if* $\ker(L + A) = \{0\}$, *then $L + A : D(L) \to Z$ is bijective and, for each L-compact mapping $G : E \to Z$, the mapping $(L+A)^{-1}G : E \to X$ is L-compact on E.*

Proof. Let $C \in \mathcal{F}(L)$. We have, on $D(L)$,

$$L + A = (L+C)[I + (L+C)^{-1}(A-C)],$$

and $(L+C)^{-1}(A-C)$ is linear and completely continuous on X. From the relation

$$\ker[I + (L+C)^{-1}(A-C)] = \ker(L+A) = \{0\},$$

and Riesz theory of linear completely continuous perturbations of identity, it follows that $I+(L+C)^{-1}(A-C)$ is a linear homeomorphism of X onto X. Hence, for each $z \in Z$, the equation

$$(L+A)x = z,\ x \in D(L),$$

is equivalent to the equation

$$(L+C)[I+(L+C)^{-1}(A-C)]x = z,\ x \in D(L),$$

and has the unique solution

$$x = [I+(L+C)^{-1}(A-C)]^{-1}(L+C)^{-1}z,$$

which shows that $(L+A)(D(L)) = Z$. If now $G : E \to Z$ is L-compact, then, from the relation

$$(L+C)^{-1}G = (L+C)^{-1}(L+A)(L+A)^{-1}G =$$

$$(L+C)^{-1}(L+C+A-C)(L+A)^{-1}G = [I+(L+C)^{-1}(A-C)]^{-1}(L+C)^{-1}G,$$

we get

$$(L+A)^{-1}G = [I+(L+C)^{-1}(A-C)]^{-1}(L+C)^{-1}G,$$

and the compactness of $(L+A)^{-1}G$ on E follows immediately. ■

2.3 An axiomatic degree theory for L-compact perturbations of a Fredholm mapping of index zero L

Let X and Z be real normed vector spaces, $L : D(L) \subset X \to Z$ a Fredholm linear mapping with zero index. Let us denote by C_L the set of couples (F, Ω) where the mapping $F : D(L) \cap \bar{\Omega} \to Z$ has the form $F = L + N$ with $N : \bar{\Omega} \to Z$ L-compact and Ω is an open bounded subset of X, satisfying the condition

$$0 \notin F(D(L) \cap \partial\Omega). \tag{3}$$

A mapping D_L from C_L into $\mathbf{Z}$ will be called a *degree relatively to* L if it is not identically zero and satisfies the following axioms.

1. **Addition-excision property**. If $(F, \bar{\Omega}) \in C_L$ and Ω_1 and Ω_2 are disjoint open subsets in Ω such that

$$0 \notin F[D(L) \cap (\bar{\Omega} \setminus (\Omega_1 \cap \Omega_2))],$$

then (F, Ω_1) and (F, Ω_2) belong to C_L and

$$D_L(F, \Omega) = D_L(F, \Omega_1) + D_L(F, \Omega_2).$$

2. **Homotopy invariance property**. If Γ is open and bounded in $X \times [0,1]$, $\mathcal{H} : (D(L) \cap \bar{\Gamma}) \times [0,1] \to Z$ has the form

$$\mathcal{H}(x, \lambda) = Lx + \mathcal{N}(x, \lambda),$$

where $\mathcal{N} : \bar{\Gamma} \to Z$ is L-compact on $\bar{\Gamma}$, and if

$$\mathcal{H}(x,\lambda) \neq 0$$

for each $x \in (D(L) \cap (\partial\Gamma)_\lambda$ and each $\lambda \in [0,1]$, where

$$(\partial\Gamma)_\lambda = \{x \in X : (x,\lambda) \in \partial\Gamma\}$$

then the mapping $\lambda \mapsto D_L(\mathcal{H}(.,\lambda),\Gamma_\lambda)$ is constant on $[0,1]$, where Γ_λ denotes the set

$$\{x \in X : (x,\lambda) \in \Gamma\}.$$

The following properties are simple consequences of the axioms.

Proposition 2.2. *(Excision property). If $(F,\Omega) \in C_L$ and if $\Omega_1 \subset \Omega$ is an open set such that*

$$0 \notin F(D(L) \cap (\bar{\Omega} \setminus \Omega_1)),$$

then $(F,\Omega_1) \in C_L$ and $D_L(F,\Omega) = D_L(F,\Omega_1)$.

Proposition 2.3. *(Existence property). If $(F,\Omega) \in C_L$ is such that $D_L(F,\Omega) \neq 0$, then F has at least one zero in Ω.*

Proposition 2.4. *(Boundary value dependence). If $(F,\Omega) \in C_L$ and $(G,\Omega) \in C_L$ are such that $Fx = Gx$ for each $x \in D(L) \cap \partial\Omega$, then $D_L(F,\Omega) = D_L(G,\Omega)$.*

A degree D_L will be said to be *normalized* if it satisfies the following third axiom.

3. **Normalization property**. If $(F,\Omega) \in C_L$, with F the restriction to $\bar{\Omega}$ of a linear one-to-one mapping from $D(L)$ into Z, then $D_L(F-b,\Omega) = 0$ if $b \notin F(D(L) \cap \Omega)$ and $|D_L(F-b,\Omega)| = 1$ if $b \in F(D(L) \cap \Omega)$.

A mapping degree D_0 was first constructed by KRONECKER [65] in 1869 when $X = Z = \mathbf{R}^n$, F is of class C^1, Ω has a regular boundary and $0 \notin F(\partial\Omega)$, and then by BROUWER [8] in 1912 when X and Z are finite dimensional oriented vector spaces, F is continuous and $0 \notin F(\partial\Omega)$ (it is called the *Brouwer degree* and usually denoted by $\deg(F,\Omega,0)$). In 1934, LERAY and SCHAUDER [76] constructed a mapping degree D_I when $X = Z$ is a Banach space, $F = I + N$ with $N : \bar{\Omega} \to X$ compact and $0 \notin (I+N)(\partial\Omega)$ (it is called the *Leray-Schauder degree* and, as it reduces to the Brouwer degree when X is finite-dimensional, it will also be denoted by $\deg(F,\Omega,0)$). Those degree mappings satisfy the three axioms above and further properties that we will use freely in the sequel. One can refer to [24], [77] and [141] for recent presentations of those degrees.

2.4 A construction of the degree mapping in the general case

Let X and Z be real vector normed spaces, $L : D(L) \subset X \to Z$ a linear Fredholm mapping with zero index and $(F,\Omega) \in C_L$, with $F = L + N$. The definition of D_L is based upon the Leray-Schauder degree and upon a simple lemma. Denote by $\mathcal{C}(L)$ the set of linear completely continuous mappings $A : X \to Z$ such that $\ker(L+A) = \{0\}$. By Proposition 2.1, $L + A : D(L) \to Z$ is bijective and $(L+A)^{-1}G$ is compact over $E \subset X$ whenever $G : E \to Z$ is L-compact. Furthermore, one has $\mathcal{F}(L) \subset \mathcal{C}(L)$.

Lemma 2.2. *If $A \in \mathcal{C}(L)$ and $B \in \mathcal{C}(L)$, and if we set $\Delta_{B,A} = (L+B)^{-1}(A-B)$, then $\Delta_{B,A}$ is completely continuous on X and*

$$I + (L+B)^{-1}(N-B) = (I + \Delta_{B,A})[I + (L+A)^{-1}(N-A)].$$

Proof. We have

$$I + (L+B)^{-1}(N-B) = I + (L+B)^{-1}(L+A)(L+A)^{-1}(N-A+A-B) =$$

$$I + (L+B)^{-1}(L+B+A-B)(L+A)^{-1}(N-A+A-B) =$$

$$I + [I + (L+B)^{-1}(A-B)](L+A)^{-1}(N-A) + (L+B)^{-1}(A-B) =$$

$$(I + \Delta_{B,A})[I + (L+A)^{-1}(N-A)].$$

The complete continuity of $\Delta_{B,A}$ on X follows from Proposition 2.1. ■

It is easy to check that $\ker(I + \Delta_{B,A}) = \ker(L+A) = \{0\}$, and hence, by the product formula of Leray-Schauder degree, we obtain

$$D_I(I + (L+B)^{-1}(N-B), \Omega) = D_I(I + \Delta_{B,A}, B(r)).D_I(I + (L+A)^{-1}(N-A), \Omega),$$

where $r > 0$ is arbitrary and $B(r) = \{x \in X : \|x\| < r\}$. We can now define a relation in $\mathcal{C}(L)$ by $B \sim A$ if and only if $D_I(I + \Delta_{B,A}, B(r)) = +1$. It is an equivalence relation over $\mathcal{C}(L)$. If we fix an orientation on $\ker L$ and on $\operatorname{coker} L = Z/R(L)$, we can for example define $\mathcal{C}_+(L)$ as the class containing the application A of the form $\pi_Q^{-1}\Lambda P$, where $\Lambda : \ker L \to \operatorname{coker} L$ is an orientation preserving isomorphism and π_Q is the restriction to $R(Q)$ of the canonical projection $\pi : Z \to \operatorname{coker} L$. Setting $J = \pi_Q^{-1}\Lambda : \ker L \to R(Q)$, it is easy to compute that

$$(L + JP)^{-1} = J^{-1}Q + K_{P,Q},$$

and hence

$$I + (L+JP)^{-1}(N - JP) = I - P + J^{-1}QN + K_{P,Q}N.$$

The following definition is therefore justified.

Definition 2.2. *If $(F, \Omega) \in C_L$, the* degree of F in Ω with respect to L *is defined by*

$$D_L(F, \Omega) = D_I(I + (L+A)^{-1}(N-A), \Omega) = deg(I + (L+A)^{-1}(N-A), \Omega, 0),$$

for any $A \in \mathcal{C}_+(L)$.

This degree was introduced in 1972 (see [85]) and systematic expositions were given in [48] and [89]. The definition given here, which comes from [90] or [91], is a minor modification of VOLKMANN's one in [136], which allows a unification of the approaches in [136] and that of PEJSACHOWICZ-VIGNOLI [116].

It is easy to see, using the properties of the Leray-Schauder degree, that D_L satisfies the three axiomes of the previous section and reduces to $D_I = \deg$ if $X = Z$ and $L = I$. Its computation is reduced to that of some Brouwer degree in the following interesting particular case.

Proposition 2.5. *If* $(F,\Omega) \in C_L$ *and* $F = L + G$ *with* $G(\bar{\Omega}) \subset Y$ *and* Y *a (finite dimensional) direct summand of* $R(L)$, *then* $(G|_{\ker L}, \Omega \cap \ker L) \in C_0$ *and*

$$|D_L(F,\Omega)| = |D_0(G|_{\ker L}, \Omega \cap \ker L)| \ (= |deg(G|_{\ker L}, \Omega \cap \ker L, 0)|).$$

Proof. Let us choose the projector $Q : Z \to Z$ such that

$$\ker Q = R(L), \ R(Q) = Y,$$

and take $A = JP$, where $J : \ker L \to Y$ is an isomorphism such that $A \in \mathcal{C}_+(L)$. Then, as $(L + A)^{-1} = J^{-1}Q + K_{P,Q}$, we have, using the definition of the Leray-Schauder degree,

$$D_L(F,\Omega) = D_I(I + (L + A)^{-1}(G - A), \Omega) =$$

$$D_I(I + J^{-1}G - P, \Omega) = D_0(J^{-1}G|_{\ker L}, \Omega \cap \ker L) =$$

$$\text{sign } \det J^{-1} D_0(G|_{\ker L}, \Omega \cap \ker L).$$

■

2.5 Existence theorems of Leray-Schauder type

The existence and homotopy invariance properties of the degree easily lead to interesting existence theorems. They all follow from the following result of the Leray-Schauder type.

Let X and Z be real normed vector spaces, $L : D(L) \subset X \to Z$ a linear Fredholm mapping of index zero, Γ open and bounded in $X \times [a,b]$, such that Γ_a is nonempty and bounded, where, for each $\lambda \in [a,b]$,

$$\Gamma_\lambda = \{x \in X : (x,\lambda) \in \Gamma\}.$$

Let $\mathcal{N} : \bar{\Gamma} \to Z$ be L-compact on Γ, let

$$\mathcal{S} = \{(x,\lambda) \in \bar{\Gamma} : x \in D(L) \text{ and } Lx + \mathcal{N}(x,\lambda) = 0\}$$

denote the solution set, and, for each $\lambda \in [a,b]$, let $\mathcal{S}_\lambda = \{x \in X : (x,\lambda) \in \mathcal{S}\}$.

Lemma 2.3. *If* $(L + \mathcal{N}(.,a), \Gamma_a) \in C_L$ *and* $D_L(L + \mathcal{N}(.,a), \Gamma_a) \neq 0$, *then there exists a closed connected subset* Σ *of* $\mathcal{S}$ *which connects* $\Gamma_a \times \{a\}$ *to either* $\Gamma_b \times \{b\}$ *or* $\{(x,\lambda) \in \partial\Gamma : \lambda \in]a,b[\,\}$.

In particular, we have the following

Corollary 2.1. *If* $(L + \mathcal{N}(.,a), \Gamma_a) \in C_L$ *and*
1) $D_L(L + \mathcal{N}(.,a), \Gamma_a) \neq 0$,
2) $Lx + \mathcal{N}(x,\lambda) \neq 0$ *for each* $x \in (\partial\Gamma)_\lambda$ *and each* $\lambda \in]a,b]$,
then there exists a closed connected subset Σ *of* $\mathcal{S}$ *which connects* $\Gamma_a \times \{a\}$ *to* $\Gamma_b \times \{b\}$.

Combining Lemma 2.3 with a topological lemma of KURATOWSKI-WHYBURN, one obtains also the following

Corollary 2.2. *If Γ is not necessarily bounded but Γ_a is bounded and the conditions of Lemma 2.3 hold, then there exists a closed connected subset Σ of $\mathcal{S}$ intersecting $\Gamma_a \times \{a\}$ and such that either*
a) Σ connects $\Gamma_a \times \{a\}$ to $\Gamma_b \times \{b\}$,
b) or Σ connects $\Gamma_a \times \{a\}$ to $\{(x,\lambda) \in \partial\Gamma : \lambda \in]a,b[\}$,
c) or Σ is unbounded.

A first useful consequence of Corollary 2.1 is the following existence theorem of the POINCARÉ-BOHL type.

Theorem 2.1. *Let $(H,\Omega) \in C_L$ and $F = L + N$ with $N : \bar{\Omega} \to Z$ L-compact and Ω open and bounded in X. Assume that the following conditions are satisfied.*
(i) $\lambda Fx + (1-\lambda)Hx \neq 0$ for each $(x,\lambda) \in (D(L) \cap \partial\Omega) \times]0,1[$.
(ii) $D_L(H,\Omega) \neq 0$.
Then the equation $Lx + Nx = 0$ has at least one solution in $D(L) \cap \bar{\Omega}$.

Proof. For each $x \in D(L) \cap \bar{\Omega}$ and each $\lambda \in [0,1]$, one has, if $H = L + K$ with K L-compact on Ω,

$$\lambda Fx + (1-\lambda)Hx = Lx + \lambda Nx + (1-\lambda)Kx = Lx + \mathcal{N}(x,\lambda),$$

with $\mathcal{N} : \bar{\Omega} \times [0,1] \to Z$ L-compact, as easily verified. If we take $\Gamma = \Omega \times [0,1]$ in Lemma 2.3, we have Γ bounded and hence $\mathcal{S}$ is bounded too. Thus, either the equation $Lx + Nx = 0$ has a solution in $D(L) \cap \partial\Omega$, or the set Γ satisfies all the conditions of Corollary 2.1. ∎

A first useful consequence of Theorem 2.1 is the following result, which can be found in [89] in the present form and whose various special cases or variants have been widely applied to ordinary and partial differential equations since the pioneering work of LERAY and SCHAUDER.

Theorem 2.2. *Let $F = L + N$, with $N : \bar{\Omega} \to Z$ L-compact, let $A : X \to Z$ be a linear L-completely continuous mapping and let $z \in (L + A)(D(L) \cap \Omega)$ satisfy the following conditions.*
(i) $\ker(L + A) = \{0\}$.
(ii) $Lx + (1-\lambda)(Ax - z) + \lambda Nx \neq 0$ for each $x \in D(L) \cap \partial\Omega$ and each $\lambda \in]0,1[$.
Then equation $Lx + Nx = 0$ has at least one solution in $D(L) \cap \bar{\Omega}$.

Proof. It suffices to take $H = L + A - z$ in Theorem 1 and to notice that $|D_L(L + A - z, \Omega)| = 1$. ∎

Another useful consequence of Theorem 2.1 is the following result (when $\ker L \neq \{0\}$) which comes from [89].

Theorem 2.3. *Let $F = L + N$, with $N : \bar{\Omega} \to Z$ L-compact and let $G : \bar{\Omega} \to Y$ be L-compact on $\bar{\Omega}$, with Y a direct summand of $R(L)$. Assume that the following conditions hold.*
(i) $Lx + (1-\lambda)Gx + \lambda Nx \neq 0$ for each $x \in D(L) \cap \partial\Omega$ and each $\lambda \in]0,1[$.
(ii) $Gx \neq 0$ for each $x \in \ker L \cap \partial\Omega$.

(iii) $D_0(G|_{\ker L}, \Omega \cap \ker L) \neq 0$.
Then equation $Lx + Nx = 0$ *has at least one solution in* $D(L) \cap \bar{\Omega}$.

Proof. Let $H = L + G$ and let $Q : Z \to Z$ be the continuous projector such that $R(Q) = Y$ and $\ker Q = R(L)$. Then $QG = G$ and $Hx = 0$ if and only if

$$QHx = 0, \ (I - Q)Hx = 0,$$

i.e.

$$Gx = 0, \ Lx = 0,$$

i.e.

$$Gx = 0, \ x \in \ker L.$$

Hence, by assumption (ii), $(H, \Omega) \in C_L$ and, using Proposition 2.5, we have

$$|D_L(H, \Omega)| = |D_0(G|_{\ker L}, \Omega \cap \ker L)|.$$

The result follows then from assumptions (i), (iii) and from Theorem 1. ∎

A useful consequence of Theorem 2.3 is the following one, which was first proved in [85]. Other more recent proofs or variants can be found in [43], [44], [45], [49], [61], [137].

Theorem 2.4. *Let* $F = L + N$, *with* $N : \bar{\Omega} \to Z$ *L-compact. Assume that the following conditions are satisfied.*
(i) $Lx + \lambda Nx \neq 0$ *for each* $(x, \lambda) \in [(D(L) \setminus \ker L) \cap \partial\Omega] \times \,]0, 1[$.
(ii) $Nx \notin R(L)$ *for each* $x \in \ker L \cap \partial\Omega$.
(iii) $D_0(QN|_{\ker L}, \Omega \cap \ker L) \neq 0$, *where* $Q : Z \to Z$ *is a continuous projector such that* $R(L) = \ker Q$.
Then equation $Lx + Nx = 0$ *has at least one solution in* $D(L) \cap \bar{\Omega}$.

Proof. Let us take, in Theorem 2.3, $Y = R(Q)$ and $G = QN$, which is L-compact on $\bar{\Omega}$. Assumption (ii) immediately implies that $QNx \neq 0$ for each $x \in \ker L \cap \partial\Omega$, and assumptions (ii) and (iii) of Theorem 2.3 are fulfilled. On the other hand, if $Lx + (1 - \lambda)QNx + \lambda Nx = 0$ for some $x \in D(L) \cap \partial\Omega$ and some $\lambda \in \,]0, 1[$, then, applying Q and $I - Q$ to both members of this equation, we obtain

$$QNx = 0, \ Lx + \lambda Nx = 0.$$

The first of those equations and assumption (ii) imply that $x \in (D(L) \setminus \ker L) \cap \partial\Omega$ and the second one then contradicts assumption (i). Thus, assumption (i) of Theorem 2.3 is satisfied and the existence of a solution follows. ∎

3 Computing the degree of autonomous equations in spaces of periodic functions

3.1 Introduction

This section is devoted to the proof of a result concerning the computation of the topological degree associated with the periodic solutions of fixed period ω of an arbitrary autonomous system under the unique assumption of the existence of an open bounded subset in the space of continuous ω-periodic functions whose boundary does not contain any possible ω-periodic solution of the system. The result (Theorem 3.1) essentially says that this degree is just the Brouwer degree of the equilibria of the system contained in the trace of the open set on the subspace of constant functions. Such a result was first proved in [10] for ordinary differential equations and in [11] for retarded functional differential equations using the KUPKA-SMALE theorem. The proof presented here, and which depends upon the S^1-equivariance of the associated fixed point operator introduced in Section 2.4, was first given in [4] in the more general setting of neutral functional differential equations.

Now, there are various ways to express the problem of ω-periodic solutions of an autonomous system

$$x' = g(x)$$

in terms of fixed points of operators defined in the space of continuous functions on $[0, \omega]$. Using duality theorems developed in [64], Ch. III, and [89], Ch. III, and Theorem 3.1, one can also express the Leray-Schauder degree of these operators in terms of the Brouwer degree of g, and this is the object of Corollary 3.1. Finally, similar results also hold which connect the Brouwer degree of the Poincaré operator for ω-periodic solutions to the Brouwer degree of g (Corollary 3.2).

3.2 The degree associated to an autonomous equation

Let $\omega > 0$ be fixed. $C_\omega = \{x \in C(\mathbf{R}, \mathbf{R}^n) : x(t+\omega) = x(t), t \in \mathbf{R}\}$ with the norm $\|x\|_\omega = \max_{t \in \mathbf{R}} |x(t)|$. Define the linear operator $\mathcal{L}$ in C_ω by $D(\mathcal{L}) = \{x \in C_\omega : x \text{ is of class } C^1\}$ and $(\mathcal{L}x)(t) = x'(t)(t \in \mathbf{R})$. If $g : \mathbf{R}^n \to \mathbf{R}^n$ is continuous, we shall consider the ω-periodic solutions of the autonomous differential equation

$$x'(t) = g(x(t)). \tag{4}$$

If $\mathcal{G} : C_\omega \to C_\omega$ is the continuous mapping defined by

$$(\mathcal{G}x)(t) = g(x(t)),$$

then finding the ω-periodic solutions of (4) is equivalent to solving the abstract equation

$$\mathcal{L}x = \mathcal{G}x \tag{5}$$

in $D(\mathcal{L}) \cap C_\omega$. Now, $x \in \ker \mathcal{L}$ if and only if $x \in D(\mathcal{L})$ and

$$x(t) = c \in \mathbf{R}^n \tag{6}$$

for all $t \in \mathbf{R}^n$, so that $\ker \mathcal{L} \simeq \mathbf{R}^n$. Now, if $y \in R(\mathcal{L})$, i.e. if $y \in C_\omega$ and

$$x'(t) = y(t)$$

for some $x \in D(\mathcal{L})$, we get by integration that

$$\mathcal{P}y := \frac{1}{\omega} \int_0^\omega y(t)dt = 0 \tag{7}$$

so that $R(\mathcal{L}) \subset \mathcal{N}(\mathcal{P})$, with $\mathcal{P}$ a continuous finite-dimensional projector in $\mathcal{C}_\omega$. Now, if (7) holds, then all the solutions of equation

$$x'(t) = y(t)$$

are ω-periodic and hence

$$\ker \mathcal{L} = R(\mathcal{P}).$$

Notice now that the space $\mathcal{C}_\omega = C(\mathbf{R}/\omega\mathbf{Z}, \mathbf{R}^n)$ is endowed with an S^1-action, i.e. a continuous mapping

$$S^1 \times \mathcal{C} \to \mathcal{C}, (\tau, x) \mapsto \tau \star x,$$

compatible with the group law $(+)$

$$\begin{aligned} \tau_1 \star (\tau_2 \star x) &= (\tau_1 + \tau_2) \star u \\ 0 \star x &= x \end{aligned}$$

for all $\tau_i \in S^1$, $x \in \mathcal{C}$, which is indeed a continuous representation of S^1 into the group of isometries of $\mathcal{C}_\omega$, defined by

$$(\tau \star x)(t) := x(t+\tau)\, (\tau \in [0, \omega]).$$

Now, $\tau \star x \in D(\mathcal{L})$ for all $\tau \in S^1$ if $x \in D(\mathcal{L})$ and, for all $t \in \mathbf{R}$

$$[\tau \star \mathcal{L}(x)](t) = \frac{dx}{dt}(\tau + t) = \mathcal{L}(\tau \star x)(t)$$

which shows that $\mathcal{L}$ is equivariant for this S^1-action. Similarly, for $x \in \mathcal{C}_\omega$, $t \in \mathbf{R}$ and $\tau \in S^1$,

$$\begin{aligned} [\tau \star \mathcal{G}(x)](t) &= g[x(t+\tau)] = g[(\tau \star x)(t)] = \\ &= \mathcal{G}(\tau \star x)(t) \end{aligned}$$

and $\mathcal{G}$ is equivariant for the S^1-action as well. Now, for $\mathcal{P}$ defined in (7) we have, for all $x \in \mathcal{C}_\omega$ and $\tau \in S^1$.

$$\tau \star \mathcal{P}(x) = \frac{1}{\omega}\int_0^\omega x(s)ds = \frac{1}{\omega}\int_0^\omega x(\tau + s)ds = \mathcal{P}(\tau \star x)$$

so that $\mathcal{P}$ is also equivariant, and consequently $R(\mathcal{P})$ and $\ker \mathcal{P}$ are S^1-invariant. Defining now as in Section 2 the (unique) generalized inverse $\mathcal{K} = \mathcal{K}_{\mathcal{P},\mathcal{P}}$ of $\mathcal{L}$ associated to $\mathcal{P}$ as

$$\mathcal{K} = (\mathcal{L}\,|_{D(\mathcal{L})\cap\ker(\mathcal{P})})^{-1} : R(\mathcal{L}) \to D(\mathcal{L}) \cap \ker \mathcal{P},$$

so that

$$\mathcal{L}\mathcal{K}v = v,\ v \in R(\mathcal{L})$$

we see that, as $R(\mathcal{L})$ is S^1-invariant and $\mathcal{L}$ equivariant,

$$\mathcal{L}\mathcal{K}(\tau \star v) = \tau \star v = \tau \star \mathcal{L}(\mathcal{K}v) = \mathcal{L}(\tau \star \mathcal{K}v).$$

Hence, $\mathcal{L}$ being one-to-one on $D(\mathcal{L}) \cap \ker \mathcal{P}$,

$$\mathcal{K}(\tau \star v) = \tau \star \mathcal{K}v$$

for all $\tau \in S^1$ and $v \in R(\mathcal{L})$, and $\mathcal{K}$ is equivariant. Now, (4) is equivalent to the fixed point problem in $\mathcal{C}_\omega$ (see Section 2.4)

$$x = \mathcal{P}x + [\mathcal{P} + \mathcal{K}(\mathcal{I} - \mathcal{P})]\mathcal{G}x := \mathcal{M}x \tag{8}$$

and the above discussion implies that $\mathcal{M}$ is equivariant for the S^1-action. Obviously $\mathcal{M}$ is continuous and $\mathcal{K} : R(\mathcal{L}) \to R(\mathcal{L})$ is compact by the Arzela-Ascoli theorem. We have thus proved the following

Proposition 3.1. *If $g : \mathbf{R}^n \to \mathbf{R}^n$ is continuous, then the problem of finding the ω-periodic solutions of (4) is equivalent to a fixed point problem*

$$x = \mathcal{M}(x)$$

in $\mathcal{C}_\omega$, where $\mathcal{M}$ is completely continuous and equivariant for the action of S^1.

We can now state and prove our main result.

Theorem 3.1. *Assume that $g : \mathbf{R}^n \to \mathbf{R}^n$ is continuous and that there exists an open bounded set $\Omega \subset \mathcal{C}_\omega$ such that (4) has no solution on $\partial\Omega$. Then*

$$D_{\mathcal{L}}(\mathcal{L} - \mathcal{G}, \Omega) := \deg(\mathcal{I} - \mathcal{M}, \Omega, 0) = (-1)^n \deg(g|_{\mathbf{R}^n}, \Omega \cap \mathbf{R}^n, 0)$$

where $\mathbf{R}^n$ is identified with the subspace of constant functions of $\mathcal{C}_\omega$.

Proof. Theorem 3.1 is a consequence of the following result about the Leray-Schauder degree of an S^1-equivariant map. ■

Let Fix $\mathcal{M}$ denote the set of fixed points of the operator $\mathcal{M}$.

Theorem 3.2. *Let X be a normed linear space, $\Omega \subset X$ open and bounded and $f : \overline{\Omega} \to X$ be a compact perturbation of identity such that $f^{-1}(0) \cap \partial\Omega = \phi$. Assume S^1 acts on X through linear isometries, $\Omega \subset X$ is invariant under this action and f is equivariant. Let $f^{S^1} : \overline{\Omega}^{S^1} \to X^{S^1}$ denote the restriction of f to the fixed point set $\overline{\Omega}^{S^1} = \overline{\Omega} \cap X^{S^1}$, $X^{S^1} = \{x \in X : \tau \star x = x \text{ for all } \tau \in S^1\}$. Then*

$$\deg(f, \Omega, 0) = \deg(f^{S^1}, \Omega^{S^1}, 0).$$

Theorem 3.2 will be proved in Subsection 3.2 and we show here how Theorem 3.1 follows. Here $X = \mathcal{C}_\omega$ and $f = \mathcal{I} - \mathcal{M}$ is equivariant. If $\Omega \subset \mathcal{C}_\omega$ is not invariant we replace it by $\tilde{\Omega} = \{x \in \Omega : \text{dist}(x, \Omega \cap \text{Fix } \mathcal{M}) < \epsilon\}$ with $0 < \epsilon < \text{dist}(\Omega \cap \text{Fix } \mathcal{M}, \partial\Omega)$. $\tilde{\Omega}$ is invariant since S^1 is path-connected, hence $\partial\Omega \cap \text{Fix } \mathcal{M} = \emptyset$ implies that $\Omega \cap \text{Fix } \mathcal{M}$ is invariant. We also used that S^1 acts through isometries. Since Fix $\mathcal{M} \cap \Omega \subset \tilde{\Omega} \subset \Omega$, the excision property of the Leray-Schauder degree yields

$$\deg(\mathcal{I} - \mathcal{M}, \Omega, 0) = \deg(\mathcal{I} - \mathcal{M}, \tilde{\Omega}, 0).$$

Now the fixed point set $\mathcal{C}_\omega^{S^1} = \{x \in \mathcal{C}_\omega : x(\cdot + \tau) = x(\cdot) \text{ for all } \tau \in [0,\omega]\}$ is the set of constant x in $\mathcal{C}_\omega$ and, for such a constant a, $\mathcal{P}a = a$, $\mathcal{P}\mathcal{G}a = \mathcal{G}a$, so that

$$a - \mathcal{M}(a) = -\mathcal{P}\mathcal{G}a = -g(a).$$

Consequently, as the Leray-Schauder degree in a finite dimensional spaces reduces to the Brouwer degree, we get, using Theorem 3.2 and excision,

$$\deg(\mathcal{I} - \mathcal{M}, \tilde{\Omega}, 0) = \deg((\mathcal{I} - \mathcal{M})|_{\mathcal{C}_\omega^{S^1}}, \tilde{\Omega} \cap \mathcal{C}_\omega^{S^1}, 0) =$$

$$\deg(-g|_{\mathbf{R}^n}, \tilde{\Omega} \cap \mathbf{R}^n, 0) = (-1)^n \deg(g|_{\mathbf{R}^n}, \tilde{\Omega} \cap \mathbf{R}^n, 0) =$$
$$(-1)^n \deg(g|_{\mathbf{R}^n}, \Omega \cap \mathbf{R}^n, 0)$$

as $\tilde{\Omega} \cap \mathbf{R}^n$ contains all constant ω-periodic solutions of (4) in Ω, i.e. all zeros of $g|_{\mathbf{R}^n}$. ■

Theorem 3.1 is a generalization of Lemma VI.1 in [89], where the case $g = -\nabla V$, with $V \in C^1(\mathbf{R}^m, \mathbf{R})$ and $\Omega \cap \mathbf{R}^m = B(0,r)$, $r > 0$, is treated.

From Theorem 3.1, using the duality theorems developed in [89], Ch.III, and [64], Ch. III, we can find other relations between the degree of some fixed point operators related to the ω-periodic solutions of (4) and the Brouwer degree of g. To this end, the following maps $\Phi_i : Y \to Y$, $i = 1,2,3$, are defined, where $Y = C([0,\omega], \mathbf{R}^n)$:

$$\Phi_1(x)(t) := x(\omega) + \int_0^t g(x(s))ds,$$

$$\Phi_2(x)(t) := x(0) + \int_0^\omega g(x(s))ds + \int_0^t g(x(s))ds,$$

$$\Phi_3(x)(t) := x(0) + (\omega - t)\int_0^\omega g(x(s))ds + \int_0^t g(x(s))ds.$$

All the Φ_i, $i = 1,2,3$, are completely continuous and their corresponding fixed points are exactly the ω-periodic solutions of (4). Moreover, $\Phi_3|_{\mathcal{C}_\omega} : \mathcal{C}_\omega \to \mathcal{C}_\omega$. Let $\Omega \subset Y$ be bounded and open (relatively to Y). In [89], the following equalities are proved, provided that there is no $x \in \partial\Omega$, ω-periodic solution of (4) :

$$\deg(I - \Phi_1, \Omega, 0) = \deg(I - \Phi_2, \Omega, 0) =$$

$$\deg(I - \Phi_3, \Omega, 0) = \deg((I - \Phi_3)|_{\mathcal{C}_\omega}, \Omega \cap \mathcal{C}_\omega, 0).$$

Indeed, it is sufficient to apply, respectively, Theorem III.1, Theorem III.4 and Proposition III.5 in [89], Ch.III. Related results can be found in [64], Section 28.

Now, we have:

Corollary 3.1. *Assume that there is no $x \in \mathcal{C}_\omega \cap \partial\Omega$ such that $x' = g(x)$. Then, for $i = 1,2,3$,*

$$\deg(I - \Phi_i, \Omega, 0) = (-1)^n \deg(g, \Omega \cap \mathbf{R}^n, 0). \tag{9}$$

Proof. It is sufficient to recall that, by [89], Th.III.7,

$$\deg(I - \Phi_3, \Omega, 0) = D_\mathcal{L}(\mathcal{L} - \mathcal{G}, \Omega \cap \mathcal{C}_\omega)$$

and then Theorem 3.1 can be applied. ■

In [103], the equality $\deg(I-\Phi_1,\Omega,0)=\deg(-g,\Omega\cap\mathbf{R}^n,0)$ is stated for the case when Ω is a ball and g is positively homogeneous of order 1, assuming that equation (4) does not possess non trivial periodic solutions of *any* period. Hence, Corollary 3.1 improves MUHAMADIEV's theorem in [103], Th. 5 (see Section 4.4 for a more detailed discussion).

Finally, we give an analogous result for the Poincaré map. Suppose that equation (4) defines a flow in $\mathbf{R}^n$, i.e. assume uniqueness and global existence for the solutions of the Cauchy problems associated to (4). For each $z\in\mathbf{R}^n$, we denote by $x(\cdot,z)$ the solution of (4) with $x(0,z)=z$. Thus, the *Poincaré operator* P_ω on $[0,\omega]$ is defined by

$$P_\omega z := x(\omega,z).$$

Let $G\subset\mathbf{R}^n$ be an open bounded set. Then, the following result holds.

Corollary 3.2. *Assume that $P_\omega z\neq z$ for all $z\in\partial G$. Then,*

$$\deg(I-P_\omega,G,0)=(-1)^n\deg(g,G,0). \tag{10}$$

Proof. We fix $R>0$ such that

$$R>\sup\{|x(t,z)| : 0\le t\le\omega,\ z\in\bar{G}\}.$$

Then, for $\Omega:=\{x\in Y : x(0)\in G, |x|_\infty<R\}$, we have:

$$\deg(I-\Phi_1,\Omega,0)=\deg(I-P_\omega,G,0). \tag{11}$$

Indeed, (11) can be obtained either from [64], Th. 28.5, observing that $G\subset\mathbf{R}^n$ and $\Omega\subset Y$ have a "common core" with respect to the ω-periodic boundary value problem for (4), or from [89], Th. III.11, Cor. III.12. Hence, Corollary 3.1 can be applied and the thesis follows.

Recall that in [6] and [63] the equality $\deg(I-P_\omega,G,0)=\deg(-g,G,0)$ is proved under the stronger condition that $x(t,z)\neq z$ for all $t\in(0,\omega]$ and $z\in\partial G$ (that is, assuming that all the points of ∂G are of *ω-irreversibility* [63]).

3.3 A proof of Theorem 3.2

It remains to prove Theorem 3.2. Approximate f to within $0<\epsilon<\operatorname{dist}(f(\partial\Omega),0)$ by $f_1:\overline{\Omega}\to V$ where $V\subset X$ is a finite-dimensional invariant linear subspace of X. Then define $f_2:\overline{\Omega}\cap V\to V$ by

$$f_2(x)=\int_{S^1}\xi\star f_1(\xi^{-1}\star x)\,d\xi=\frac{1}{\omega}\int_0^\omega \tau\star f_1((-\tau)\star x)\,d\tau.$$

Here we consider S^1 as the complex numbers of modulus one as we shall do it from now on. So $\tau\in\mathbf{R}/\omega\mathbf{Z}$ corresponds to $\xi=e^{2\pi i\tau/\omega}\in S^1$. It is easy ot check that f_2 is equivariant and $\|f_2-f\|\le\epsilon$ (use $f(x)=\int_{S^1}\xi\star f(\xi^{-1}\star x)\,d\xi$ and the fact that S^1 acts through isometries). Therefore $\deg(f,\Omega,0)=\deg(f_2,\Omega,0)$ and $\deg(f^{S^1},\Omega^{S^1},0)=\deg(f_2^{S^1},\Omega^{S^1},0)$. We may assume that $V\cong\mathbf{R}^k\times\mathbf{C}^r$ and S^1 acts on V via

$$\xi\star(u,z)=(u,\xi^{n_1}z_1,\cdots,\xi^{n_r}z_r)$$

where $u \in \mathbf{R}^k$, $z = (z_1, \cdots, z_r) \in \mathbf{C}^r$ and $n_1, \cdots, n_r \in \mathbf{Z}\backslash\{0\}$. In particular, $V^{S^1} \cong \mathbf{R}^k$. In the following we write $V_n \cong \mathbf{C}$ for the representation of S^1 given by $\xi \star z = \xi^n z$. Thus $V \cong \mathbf{R}^k \times V_{n_1} \times \cdots \times V_{n_r}$. Set $n = n_1 \cdot \cdots \cdot n_r$ and $m_i = n/n_i$ $(1 \leq i \leq r)$. Consider the maps

$$\varphi : \mathbf{R}^k \times V_1^r \to V, (u, z_1, \cdots, z_r) \mapsto (u, z_1^{n_1}, \cdots, z_r^{n_r})$$

and

$$\psi : V \mapsto \mathbf{R}^k \times V_n^r, (u, z_1, \cdots, z_r) \mapsto (u, z_1^{m_1}, \cdots, z_r^{m_r}).$$

φ and ψ are equivariant, hence so is $\psi \circ f_2 \circ \varphi$. By the product formula for the degree

$$\deg(\psi \circ f_2 \circ \varphi, \varphi^{-1}(\Omega \cap V), 0) =$$

$$= \deg(\psi) \cdot \deg(f_2, \Omega, 0) \cdot \deg(\varphi) = \deg(f_2, \Omega, 0) \cdot n^r,$$

since $\deg(\varphi) = n_1 \cdot \cdots \cdot n_r$ and $\deg(\psi) = m_1 \cdot \cdots \cdot m_r$. Observe that $\varphi^{-1}(0) = \{0\}$ and we wrote $\deg(\varphi)$ for the local degree of φ at 0, and similarly for ψ.

Now set $F = \psi \circ f_2 \circ \varphi$ and $\mathcal{O} = \varphi^{-1}(\Omega \cap V) \subset \mathbf{R}^k \times V_1^r$. φ and ψ induce the identity on the fixed point component $\mathbf{R}^k$. Thus $\deg(F^{S^1}, \mathcal{O}^{S^1}, 0) = \deg(f_2^{S^1}, \Omega^{S^1} \cap V, 0)$. Therefore it suffices to prove

$$\deg(F, \mathcal{O}, 0) = \deg(F^{S^1}, \mathcal{O}^{S^1}, 0) \cdot n^r.$$

To see this, choose a neighbourhood $\mathcal{U}$ of $F^{-1}(0) \cap \mathbf{R}^k$ and $\epsilon > 0$ such that $\overline{\mathcal{U} \times B_\epsilon(0)} \subset \mathcal{O}$. Define

$$F_1(u, z) = (F^{S^1}(u), z_1^n, \cdots, z_r^n)$$

for $(u, z) \in \overline{\mathcal{U} \times B_\epsilon(0)}$ and

$$F_1(u, z) = F(u, z)$$

for $(u, z) \in \mathcal{O}^{S^1} \cup \partial\mathcal{O}$. Here $F^{S^1} : \overline{\mathcal{O}} \cap \mathbf{R}^k \to \mathbf{R}^k$ denotes the restriction as usual. Observe that for $(u, z) \in \mathcal{O}^{S^1} \cap \overline{\mathcal{U} \times B_\epsilon(0)}$ we have $z = 0$ and by equivariance $F(u, 0) = (F^{S^1}(u), 0)$. Thus F_1 is well defined. Then extend F_1 continuously over all of $\overline{\mathcal{O}}$ and set

$$F_2(u, z) = \int_{S^1} \xi \star F_1(\xi^{-1} \star (u, z)) d\xi.$$

This coincides with F_1 where F_1 is equivariant, in particular in $\mathcal{O}^{S^1} \cup \partial\mathcal{O} \cup \overline{\mathcal{U} \times B_\epsilon(0)}$. Consequently $\deg(F_2, \mathcal{O}, 0) = \deg(F, \mathcal{O}, 0)$ and $\deg(F_2^{S^1}, \mathcal{O}^{S^1}, 0) = \deg(F^{S^1}, \mathcal{O}^{S^1}, 0)$. Now, for $v = (v_1, \cdots, v_r)$ with $v_i \in \mathbf{C}^r \subset \mathbf{R}^k \times \mathbf{C}^r$ we define

$$F_v(u, z) = F_2(u, z) - \sum_{i=1}^{r} z_i^n v_i.$$

One checks easily that $F_v : \overline{\mathcal{O}} \to \mathbf{R}^k \times V_n^r$ is equivariant and that the maps $(v, u, z) \mapsto F_v(u, z)$ has 0 as a regular value. Now the parametrized Sard theorem (cf. [100], Theorem 5.2) implies that there exists $v \in \mathbf{C}^{r^2}$ arbitrarily close to 0 such that 0 is a regular value of F_v. If v is small enough the degrees of F_v and F_2 are the same. Also, restricted to $\mathbf{R}^k$, F_v and F coincide. But if 0 is a regular value of F_v, then $F_v^{-1}(0) \subset \mathbf{R}^k$. This is obvious since any $(u, z) \in F_v^{-1}(0)$ with $z \neq 0$ yields a one-dimensional manifold $\{\xi \star (u, z) : \xi \in S^1\}$

of zeroes contradicting the regularity condition. So $F_v^{-1}(0) = F^{-1}(0) \cap \mathbb{R}^k$. Now, in the neighbourhood $\overline{\mathcal{U} \times B_\epsilon(0)}$ of $F_v^{-1}(0)$ we have by construction

$$F_v(u;z) = (F(u), z_1^n, \cdots, z_r^n) - \sum_{i=1}^{r} z_i^r v_i.$$

Applying excision, homotopy invariance and the product formula one more we obtain

$$\deg(F, \mathcal{O}, 0) = \deg(F_v, \mathcal{O}, 0) = \deg(F^{S^1}, \mathcal{O}^{S^1}, 0) \cdot n^r.$$

■

We conclude the subsection by some remarks.

Remark 3.1. Theorem 3.2 generalizes to the fixed point index. Let $G = S^1$ act on the normed linear space X through isometries and let $A \subset X$ be a G-ANR. Let $C : \Omega \to A$ be a continuous equivariant map defined on an open invariant subset Ω of A. Suppose that Fix $C = \{x \in A : C(x) = x\}$ is compact and that C is compact one some open neighbourhood of Fix C. Then

$$\mathrm{ind}(C, \Omega) = \mathrm{ind}(C^G, \Omega^G)$$

where ind denotes the fixed point index.

Remark 3.2. A slight modification of the ideas used to prove Theorem 3.2 yields the following result. Let

$$f : \mathcal{O} \subset \mathbb{R}^k \times V_{m_1} \times \cdots \times V_{m_r} \to \mathbb{R}^k \times V_{n_1} \times \cdots \times V_{n_r}$$

be an equivariant map where $n_i, m_i \subset \mathbb{Z}\backslash\{0\}$ and S^1 acts on V_n as above. If $\mathcal{O}$ is open, bounded, invariant and $f^{-1}(0) \cap \partial\mathcal{O} = \phi$, then

$$\deg(f, \mathcal{O}, 0) \cdot m_1 \cdot \cdots \cdot m_r = \deg(f^{S^1}, \mathcal{O}^{S^1}, 0) \cdot n_1 \cdot \cdots \cdot n_r.$$

If $k = 0$, then either $\mathcal{O}^{S^1} = \{0\}$ or $\mathcal{O}^{S^1} = \phi$. This implies $\deg(f^{S^1}, \mathcal{O}^{S^1}, 0) = 1$ or $\deg(f^{S^1}, \mathcal{O}^{S^1}, 0) = 0$, respectively. In the case $\mathcal{O}^{S^1} = \{0\}$ and $m_i = n_i$ we recover Theorem 5.4 of [100]. Thus the degree of an S^1-equivariant map f is completely determined by the degree of its fixed point part f^{S^1} and by the action of S^1 on the domain and the range. The behavior of f outside of the set of points fixed by the action is irrelevant. Also, if $\deg(f^{S^1}, \mathcal{O}^{S^1}, 0) = 1$, a map f as above can only exist if $m_1 \cdot \cdots \cdot m_r$ divides $n_1 \cdot \cdots \cdot n_r$.

Remark 3.3. The results on the degree of S^1-equivariant maps are of course well known, and the proof of our Theorem 3.2 and Remark 3.3 is related to that given for a special case by NIRENBERG in Theorem 3 of [106] and reproduced in [100], Theorem 5.4. We included a proof since our situation is more general than the one considered in [106] and [100] and since, in most other papers, still more general situations have been considered and correspondingly the proofs are more complicated. We refer the reader to [131], Section II.5.a for a survey of related results and references to the literature, and to the recent papers [30], [118] and [138].

4 Continuation theorems for periodic perturbations of nonlinear autonomous systems

4.1 Introduction

In this subsection we shall deal with the periodic boundary value problem (BVP)

$$x' = f(t, x), \tag{12}$$

$$x(0) = x(\omega), \tag{13}$$

where $f : [0, \omega] \times \mathbf{R}^n \to \mathbf{R}^n$ is a Caratheodory function ($\omega > 0$). We recall that if $f : \mathbf{R} \times \mathbf{R}^n \to \mathbf{R}^n$ is ω-periodic in the first variable, then any solution of (12) – (13) can be extended to a classical (i.e. absolutely continuous) ω-periodic solution of (12) defined on the whole real line. Accordingly, we call in what follows ω-periodic any solution of (12) satisfying (13). We define $\mathcal{C}_\omega$ by $\{x \in C([0, \omega], \mathbf{R}^n) : x(0) = x(\omega)\}$.

The periodic BVP plays a central role in the theory of ordinary differential equations for its significance in several applications (see [63],[122],[124]). Many authors have treated problem (12) – (13) by means of topological methods; in such a framework, continuation theorems of Leray-Schauder's type turn out to be specially suitable for the existence problem. Basically, the 'continuation' is performed through an admissible homotopy carrying the given problem to a simpler one. This simpler one may be an autonomous equation whose ω-periodic solutions consist in an odd number of nondegenerate equilibria, like in STOPPELLI's pioneering work [132], or a linear equation having only the trivial ω-periodic solution (see e.g. [89], references for Theorem IV.5) or an odd differential equation. By the fundamental properties of topological degree theory, such an approach will only succeed in problems having an odd degree.

Two different methods have been developed by KRASNOSEL'SKII [63] and the author [89] which carry the problem to simpler ones induced by some associated autonomous vector fields. On one side, in [63] a homotopy through the trajectories of (12) is considered and the existence of a fixed point of the Poincaré map is proved via a degree theoretic assumption on the autonomous 'freezed' vector field $f(0, \cdot) : \mathbf{R}^n \to \mathbf{R}^n$. In this case, the admissibility of the homotopy is guaranteed by the ω-*irreversibility* condition defined above. On the other hand, in [81], system (12) is embedded into the parametrized family of equations $x' = \lambda f(t, x)$, $\lambda \in]0, 1[$, and the solvability of (12) – (13) is ensured via a degree theoretic assumption on the autonomous 'averaged' vector field $\bar{f} : z \longmapsto (1/\omega) \int_0^\omega f(s, z)\, ds$. In this case another transversality condition is required for the admissibility of the homotopy : for each $\lambda \in]0, 1[$, any ω-periodic solution x of the equation $x' = \lambda f(t, x)$ such that $x(t) \in \bar{G}$ for all $t \in [0, \omega]$ satisfies $x(t) \in G$ for all $t \in [0, \omega]$. In the special case where f is autonomous, those conditions essentially lead to the assumption that no periodic solution of *any* period belonging to $]0, \omega]$ starts from ∂G.

Those two theorems have found useful applications in the literature (see for instance the references in [5],[48],[64],[89],[109],[124],[126]). For other different but related results see [70],[121],[130].

An important situation which occurs in several applications corresponds to the case when the non-autonomous field $f(t,x)$ splits as

$$f(t,x) := g(x) + e(t,x), \tag{14}$$

where $e(\cdot,\cdot)$ satisfies suitable growth conditions (e.g. is bounded). In such a situation, it is natural to choose the homotopy field $h(t,x;\lambda) := g(x) + \lambda e(t,x)$, $\lambda \in [0,1]$ and to assume that the equation $x' = g(x)$ has no ω-periodic solution in $\bar{G}$ starting from ∂G; with such a weak assumption, none of the previously quoted continuation theorems can be directly applied. The degree computations of Section 3 will provide a new continuation result for (12) – (13) which is particularly suitable for dealing with nonlinearities like (14) under weaker assumptions that those of the previously quoted existence theorems. To do this, we assume that

$$f(t,x) := h(t,x;1),$$

where $h : [0,\omega] \times \mathbf{R}^n \times [0,1] \to \mathbf{R}^n$ is a Caratheodory function such that for $\lambda = 0$ the map h is autonomous, i.e. there is a continuous function $g : \mathbf{R}^n \to \mathbf{R}^n$ such that

$$g(x) = h(t,x;0),$$

for almost all $t \in [0,\omega]$ and all $x \in \mathbf{R}^n$. By the results of Section 3, the computation of the degree associated to this system will be again reduced to the computation of the Brouwer degree associated to g.

A first class of applications are of perturbational type, i.e. require $|e|_\infty$ to be sufficiently small. In this case, the assumptions upon g are rather mild and the results generalize in various ways earlier contributions of Amel'kin-Gaishun-Ladis [1], Berstein-Halanay [6], Cronin ([18],[19],[20]), Gomory [50], Halanay ([56],[58]), Hale-Somolinos [59], Lando ([67],[68]), Pliss [117], Srzednicki [130] and Ward [139]. A second class of results, of global type, deals with the case where g is positively homogeneous of some order and the corresponding results improve in various directions earlier contributions of Dancer [22], Fonda-Habets [40], Fonda-Zanolin [39], Fučik [41], Krasnosel'skii Zabreiko [64], Lazer-McKenna [71], Lasota [69] and Muhamadiev ([102],[103]). More results can be found in [10], which contains in particular continuation theorems for periodic solutions of differential equations on Euclidian neighbourhood retracts which complement those of [15] and [16]. In this case, the Leray-Schauder degree has to be replaced by some index for mappings in absolute neighbourhood retracts (see e.g. [53]) and the Brouwer degree by the index of the rest points (see [129]).

4.2 The continuation theorems

We now state and prove some continuation theorems. Recall that

$$h(t,x;0) = g(x), \qquad h(t,x;1) = f(t,x).$$

First, we give our main result for the solvability of

$$x' = f(t,x), \tag{15}$$

$$x(0) = x(\omega). \tag{16}$$

Theorem 4.1. *Let $\Omega \subset \mathcal{C}_\omega$ be an open bounded set such that the following conditions are satisfied:*

(p_1) *there is no $x \in \partial\Omega$ such that*

$$x' = h(t, x; \lambda), \qquad \lambda \in [0, 1[; \tag{17}$$

(p_2)

$$\deg(g, \Omega \cap \mathbf{R}^n, 0) \neq 0.$$

Then (15) – (16) *has at least one solution $x \in \bar{\Omega}$.*

Proof. We use the framework of coincidence degree theory as in Theorem 3.1. Beside the spaces and the operators considered there, we further define $\mathcal{H} : \mathcal{C}_\omega \times [0, 1] \to L_1(0, \omega; \mathbf{R}^n)$ by

$$\mathcal{H}(x; \lambda)(t) := h(t, x(t); \lambda).$$

According to [89]. Ch. VI, $\mathcal{H}$ is L-compact on $\bar{\Omega} \times [0, 1]$. We remark that x is a solution of $x' = h(t, x; \lambda)$, $\lambda \in [0, 1]$, with $x(0) = x(\omega)$, if and only if $x \in D(\mathcal{L})$ is a solution of the coincidence equation $\mathcal{L}x = \mathcal{H}(x; \lambda)$, $\lambda \in [0, 1]$. In particular, (15) – (16) is equivalent to $\mathcal{L}x = \mathcal{H}(x; 1)$.

Without loss of generality, we suppose that (p_1) holds for $\lambda \in [0, 1]$ in (15). Otherwise, the result is proved for $x \in \partial\Omega$. Accordingly, by the definition of $\mathcal{H}(\cdot; \lambda)$ and using (p_1) we have:

$$\mathcal{L}x \neq \mathcal{H}(x; \lambda), \qquad \lambda \in [0, 1],$$

for all $x \in D(\mathcal{L}) \cap \partial\Omega$. Thus, we can apply the homotopy property of the coincidence degree and obtain:

$$D_{\mathcal{L}}(\mathcal{L} - \mathcal{H}(., 0), \Omega) = D_{\mathcal{L}}(\mathcal{L} - \mathcal{H}(\cdot; 1), \Omega). \tag{18}$$

Assumption (p_1) (for $\lambda = 0$) ensures that Theorem 3.1 can be applied, so that (18) and (p_2) imply :

$$|D_{\mathcal{L}}(\mathcal{L} - \mathcal{H}(\cdot; 1), \Omega)| = |\deg(g, \Omega \cap \mathbf{R}^n, 0)| \neq 0.$$

Hence, by the existence property of the coincidence degree, there is $\tilde{x} \in D(\mathcal{L}) \cap \Omega$ such that $\mathcal{L}\tilde{x} = \mathcal{H}(\tilde{x}; 1)$; thus $\tilde{x}$ is a solution to (15) – (16). ■

An immediate consequence of Theorem 4.1 based on a suitable choice of the set Ω is the following one.

Corollary 4.1. *Let G be a bounded open subset of $\mathbf{R}^n$. Suppose that the following conditions are satisfied :*

(j_1) *('bound set' condition) : for any solution x of* (17) – (16) *such that $x(t) \in \bar{G}$ for all $t \in [0, \omega]$, one has $x(t) \in G$ for all $t \in [0, \omega]$;*

(j_2)

$$\deg(g, G, 0) \neq 0.$$

Then (15) – (16) *has at least one solution x such that $x(t) \in \bar{G}$, for all $t \in [0, \omega]$.*

Proof. It is sufficient to define (in the setting of Theorem 4.1):

$$\Omega := \{x \in \mathcal{C}_\omega : \forall t \in [0, \omega], x(t) \in G\}$$

and to check that (p_1) and (p_2) are fulfilled. For brevity, we omit the details.

Remark 4.1. Corollary 4.1 is a continuation theorem analogous to that of [89]. Namely, in [89] the bound set condition is required for the equations

$$x' = \lambda h(t, x; \lambda), \quad \lambda \in]0, 1[, \tag{19}$$

with $h(t, x; 1) = f(t, x)$, and, in place of (j_2), the Brouwer degree of the averaged vector field $\bar{h}_0(z) := (1/\omega) \int_0^\omega h(s, z; 0)ds$ is considered.

A comparison between this continuation theorem and Corollary 4.1 can be made by means of the following examples.

Example 4.1. Let us consider the planar system

$$x_1' = x_2, \qquad x_2' = -\mu x_1^+ + \nu x_1^- + p(t),$$

with $p \in L^1([0,\omega], \mathbf{R})$, $\mu > 0$, $\nu > 0$, $x^+ := \max\{x, 0\}$, $x^- := \max\{-x, 0\}$, which comes from the study of the equivalent second order scalar equation $x'' + \mu x^+ - \nu x^- = p(t)$.

It is easy to prove that Corollary 4.1 can be applied with

$$h(t, x; \lambda) := (x_2, -\mu x_1^+ + \nu x_1^- + \lambda p(t)), \ \lambda \in [0, 1]$$

and $G = B(0, R)$, for $R > 0$ sufficiently large, provided that

$$n(\mu^{-1/2} + \nu^{-1/2}) \neq \omega/\pi, \qquad \text{for every } n \in \mathbf{N}. \tag{20}$$

Indeed, in this case a priori bounds for the ω-periodic solutions are available (see [22],[41]). On the other hand, if we consider the system corresponding to (19),

$$x_1' = \lambda x_2, \qquad x_2' = \lambda\big(-\mu x_1^+ + \nu x_1^- + p(t)\big), \qquad \lambda \in]0, 1],$$

the a priori bounds for the ω-periodic solutions can be found only if

$$\lambda^{-1} n(\mu^{-1/2} + \nu^{-1/2}) \neq \omega/\pi, \qquad \text{for every } n \in \mathbf{N} \text{ and } \lambda \in]0, 1]. \tag{21}$$

Note that (21) holds if and only if $(\mu^{-1/2} + \nu^{-1/2}) > \omega/\pi$.

Hence, it is easy to choose μ and ν such that (20) holds, while (21) does not. This elementary example shows that there are situations in which Theorem 4.1 may be more directly used.

Example 4.1 deals with a periodically perturbed autonomous system. In the case of a general non-autonomous equation

$$x' = f(t, x),$$

a natural choice for the homotopy in applying Theorem 4.1 is to take

$$h(t, x; \lambda) = (1 - \lambda)\bar{f}(x) + \lambda f(t, x),$$

where $\bar{f}$ is the averaged vector field defined by

$$\bar{f}(x) = (1/\omega) \int_0^\omega f(s, x)ds.$$

Another consequence of Corollary 4.1 can be deduced in the case of planar systems ($n = 2$) for which equation (17) takes the form

$$x_1' = x_2 - \lambda g_1(x_1) + \lambda P(t), \qquad x_2' = -g_2(t, x_1; \lambda), \tag{22}$$

where $g_1 : \mathbf{R} \to \mathbf{R}$ and $P : [0, \omega] \to \mathbf{R}$ are continuous functions and $g_2 : [0, \omega] \times \mathbf{R} \times [0, 1] \to \mathbf{R}$ satisfies the Caratheodory conditions.

Systems like (22) come in a natural way from the study of the parametrized scalar Liénard equation

$$x'' + \lambda \psi_1(x) x' + \psi_2(t, x; \lambda) = \lambda p(t),$$

imposing $g_1(x_1) := \int_0^{x_1} \psi_1(s) ds$, $g_2 := \psi_2$, $P(t) := \int_0^t p(s) ds$ (usually, $\int_0^\omega p = 0$ is also assumed in order to get $P(0) = P(\omega)$). In this particular situation, the following one-sided continuation theorem can be proved.

Corollary 4.2. *Suppose that* $g_2(t, z; 0) := g_2(z)$ *and assume that there are constants* $R \geq d > 0$ *such that*

$$g_2(t, z; \lambda) \cdot z > 0, \qquad \text{for a.e. } t \in [0, \omega] \quad \text{and all } \lambda \in [0, 1[, \ |z| \geq d$$

and

$$\max\{x_1(t) : t \in [0, \omega]\} \neq R,$$

for any solution (x_1, x_2) *of* (22) – (16), *with* $\lambda \in [0, 1[$. *Then system* (22) *has at least one* ω*-periodic solution for* $\lambda = 1$.

The proof of Corollary 4.2 can be performed through the construction of an open rectangle $G =]-M, R[\times]-M, M[\subset \mathbf{R}^2$ such that condition (j_1) of Corollary 4.1 is satisfied with respect to the solutions of (22) – (16). The choice of the constant $M \geq R$ follows by the estimates developed in [99] and [108]. We omit the rest of the proof referring to these papers for the needed computations. We note that Corollary 4.2 (or some slight variants of it) is the basic tool for the proof of some recent results concerning the periodic BVP for some Liénard and Duffing equations under one-sided growth restrictions on the restoring term ψ_2 (see [29],[37]). Our result also improves [30], Lemma 1.

Remark 4.2. We point out that the results of this section may be extended to the periodic BVP for n-th order differential systems :

$$x^{(m)} + F(t, x, x', ..., x^{(m-1)}) = 0, \tag{23}$$

$$x^{(i)}(0) = x^{(i)}(\omega), \qquad i = 0, 1, ..., m-1, \tag{24}$$

with $F : [0, \omega] \times \mathbf{R}^{mn} \to \mathbf{R}^n$, by means of the standard reduction of (23) – (24) to the periodic BVP for a system of m first order equations in $\mathbf{R}^n$.

More precisely, we assume that there are $H : [0, \omega] \times \mathbf{R}^{mn} \times [0, 1] \to \mathbf{R}^n$, which fulfil the Caratheodory assumptions and $G : \mathbf{R}^{mn} \to \mathbf{R}^n$, such that

$$F(t, x, x', ..., x^{(m-1)}) = -H(t, x, x', ..., x^{(m-1)}; 1),$$

$$G(x, x', ..., x^{(m-1)}) = -F(t, x, x', ..., x^{(n-1)}; 0).$$

We also define $g : \mathbf{R}^n \to \mathbf{R}^n$, by

$$g(z) := G(z, 0, ..., 0), \qquad z \in \mathbf{R}^n.$$

Then, we have the following result.

Corollary 4.3. *Assume that there is $R > 0$ such that*

$$\max\{|x_i|_\infty : i = 1, ..., n-1\} < R,$$

for all possible solutions x of

$$x^{(m)} = H(t, x, x', ..., x^{(m-1)}; \lambda), \qquad \lambda \in [0, 1[,$$

satisfying the boundary condition (24). *Suppose that, for* $r \geq R$,

$$\deg(g, B(0, r), 0) \neq 0.$$

Then (23) – (24) *has at least one solution.*

The proof follows straightforwardly from Corollary 4.2, arguing like in [81], and therefore it is omitted.

We recall that in [23], p. 677 a similar result has been obtained for a second order scalar equation using a different approach based upon some equivariant degree theory.

As a final result, we give a continuation theorem based on the study of the Poincaré map.

For each $z \in \mathbf{R}^n$, $\lambda \in [0, 1]$, we denote by $x(\cdot, z; \lambda)$ the solution of $x' = h(t, x; \lambda)$ such that $x(0, z; \lambda) = z$. As usual, to do this, we assume uniqueness and global existence for the solutions of the Cauchy problems associated to (17). The Poincaré operator $P_{\omega,\lambda} : \mathbf{R}^n \to \mathbf{R}^n$ is defined by

$$P_{\omega,\lambda}(z) = x(\omega, z; \lambda).$$

Then, we have the following result.

Theorem 4.2. *Let $G \subset \mathbf{R}^n$ be open and bounded. Assume that the following conditions are satisfied:*

(m_1) $$P_{\omega,\lambda}(z) \neq z \quad \text{for all } z \in \partial G, \quad \lambda \in [0, 1[;$$

(m_2) $$\deg(g, G, 0) \neq 0.$$

Then (15) – (16) *has at least one solution.*

Proof. Without restriction, we can suppose that (m_1) holds with $\lambda \in [0, 1]$. Then, it is sufficient to observe that assumption (m_1) ensures that the map $(I - P_{\omega,\lambda})$ is an admissible homotopy, so that, by the homotopy invariance of the Brouwer degree.

$$\deg(I - P_{\omega,1}, G, 0) = \deg(I - P_{\omega,0}, G, 0).$$

Furthermore, Corollary 3.2 is applicable, so that

$$\deg(I - P_{\omega,1}, G, 0) = (-1)^n \deg(g, G, 0).$$

Hence, there is $z \in \bar{G}$ such that $P_{\omega,1}(z) = z$. ∎

4.3 Applications to small perturbations of autonomous systems

In this section and the following one, we deal with the problem of the existence of solutions x to

$$x' = f(t,x), \tag{25}$$

$$x(0) = x(\omega), \tag{26}$$

such that $x(t) \in \bar{G}$ for all $t \in [0,\omega]$, where G is an open bounded subset of $\mathbf{R}^n$.

We state some consequences of Theorem 4.1 and of its corollaries.

Throughout this section and the next one, we assume that the nonlinear field f splits as

$$f(t,x) = g(x) + e(t,x), \tag{27}$$

where the function $g : \mathbf{R}^n \to \mathbf{R}^n$ is continuous and $e : [0,\omega] \times \mathbf{R}^n \to \mathbf{R}^n$ satisfies the Caratheodory assumptions. We first consider the case of 'small perturbations'.

Corollary 4.4. *Assume that the following conditions are satisfied:*
(k_1) *for any ω-periodic solution x of*

$$x' = g(x) \tag{28}$$

such that $x(t) \in \bar{G}$ for all $t \in [0,\omega]$, one has $x(t) \in G$ for all $t \in [0,\omega]$;

(k_2) $$\deg(g, G, 0) \neq 0.$$

Then there is $\varepsilon_0 > 0$ such that, for any forcing term $e(\cdot,\cdot)$ with $|e(\cdot,z)|_\infty \leq \varepsilon_0$ for all $z \in \bar{G}$, system (25) *has at least one ω-periodic solution x such that $x(t) \in \bar{G}$ for all $t \in [0,\omega]$.*

Proof. We apply Corollary 4.1. We imbed (25) in the family of parametrized equations

$$x' = h(t,x;\lambda) := g(x) + \lambda e(t,x), \qquad \lambda \in [0,1], \tag{29}$$

and we claim that there is $\varepsilon_0 > 0$ such that for every function $e(\cdot,z)$ with $|e(\cdot,z)|_\infty \leq \varepsilon_0$ for all $z \in \bar{G}$, the set G is a "bound set" for (25).

Indeed, assume by contradiction that, for each $n \in \mathbf{N}$, there is a function e_n such that $|e(\cdot,z)|_\infty \leq 1/n$ for all $z \in \bar{G}$ and there is an ω-periodic function x_n satisfying

$$x_n' = g(x_n) + \lambda_n e_n(t,x_n), \qquad \lambda_n \in [0,1], \tag{30}$$

such that $x_n(t) \in \bar{G}$ for all t and $x_n(t_n) \in \partial G$ for some $t_n \in [0,\omega]$. By Ascoli-Arzelà's theorem we have that there is an ω-periodic solution x^* of (28), with $x^*(t) \in \bar{G}$ forall $t \in \mathbf{R}$, such that (up to subsequences) $x_n \to x^*$ uniformly on $[0,\omega]$. Moreover, for $t_n \to t^*$, we have $x^*(t^*) \in \partial G$. Thus a contradiction with (k_1) is reached and the claim is proved, so that (j_1) is satisfied for $e(\cdot,\cdot)$ sufficiently small. ∎

With elementary changes in the proof it can be seen that the result is still true when $e(t,x) = e(t)$ and bounds for $|e|_1$ are considered.

Corollary 4.4 enables us to recover a number of previous results, thanks especially to the rather weak condition (k_1). For example, (k_1) is satisfied whenever the flow π^0 induced by (28) is *dissipative* (i.e. there is a compact set $K \subset \mathbb{R}^n$ such that for each $x \in \mathbb{R}^n$ there is $t_x \geq 0$ with $\pi^0(t,x) \in K$ for all $t \geq t_x$); indeed, if this is the case, then $\deg(g,G,0) = (-1)^n \chi(\mathbb{R}^n) = (-1)^n$, for every $G \supset K$, where χ is the *Euler-Poincaré characteristic* (see [64], [129], Th.6.1). Hence, Corollary 4.4 guarantees the existence of periodic solutions for small periodic perturbations of autonomous dissipative systems. In this manner, we recover some classical results contained in [20],[59],[117].

Now, we discuss other results for the existence of solutions to (25) – (26) where some conditions less general than (k_1) are required.

In the two-dimensional case, CRONIN ([18],[19],[20]) and LANDO ([67],[68]) deal with periodic perturbations of autonomous systems of the form:

$$x' = X(x,y) + \varepsilon E_1(t), \qquad y' = Y(x,y) + \varepsilon E_2(t). \tag{31}$$

Following GOMORY's approach [50], the authors are led to construct a simple closed curve J (containing the origin in its interior) such that the unperturbed system

$$x' = X(x,y), \qquad y' = Y(x,y) \tag{32}$$

has no closed orbits intersecting J.

Clearly, in this situation (k_1) is satisfied and condition (k_2) either is explicitly required (see [67],[68]), or it is an implicit consequence of other hypotheses. For instance, in [18],[19], [20] it is assumed that "the point at infinity is strongly stable relative to (32)". However, in this case it can be proved that $\deg((X,Y),B(0,R),0) = 1$, for R sufficiently large. From the above discussion, it follows that Corollary 4.4 contains all the results proved in [18], Th.2, [19], Th. 6, [20], Th. 2, [67], Th. 3, [68], Th. 3.

On the other hand, we observe that none of the above quoted theorems is suitable for dealing with systems like

$$x' = -y^3 + \varepsilon E_1(t), \qquad y' = x^3 + \varepsilon E_2(t) \tag{33}$$

(see [19], p. 159]). while Corollary 4.4 applies.

We also note that many regularity hypotheses which are required in [18],[19], [20],[67],[68] are avoided using our approach.

Another condition (stronger than $(k_1) \wedge (k_2)$) leading to the existence of ω-periodic solutions of (31) was given in [1],Th. 2, where it is assumed that the origin is an isolated critical point with non-zero index and is not an isochronous center of period $\omega/k \quad (k \in \mathbb{N})$. In fact, in this case it is sufficient to take $G = B(0,\delta)$, with $\delta > 0$ sufficiently small. On the same line, see [6], Th. 2. Finally, we mention that, by means of Corollary 4.4, we can give an easy proof of the

Nemitzkii conjecture. *(first settled by* HALANAY *[56]). If the autonomous system* (32) *has a limit cycle, then there is least one ω-periodic solution of* (31), *for* ε *sufficiently small.*

Again, Corollary 4.4 may be applied, choosing G such that ∂G is 'sufficiently close' to the limit cycle.

In the higher dimensional case, Corollary 4.4 is an improvement of [1], Th. 1, [6], Th. 1, [58], Th. 3.13, where, besides (k_2), various specific conditions are required, such as, e.g. [58],

'The origin is the only critical point of (28) in a neighbourhood G of itself, and (28) has no periodic solutions of period ω', $0 < \omega' \leq \omega$, passing through points of ∂G'.

Corollary 4.4 is also a generalization of [130], Th. 4, [140], Th. 4.1, (c_1)], where, instead of (k_1), the existence of a compact isolating neighbourhood K for the flow induced by (28) is required.

Indeed, if this is the case, then $G = \text{int}\, K$ is suitable for the validity of Corollary 4.4. Again, equation (33) provides an example of applicability of our result while [130],[140] cannot be used (see also Example 3 below).

We end this subsection with an example of a system which is non-dissipative and such that, furthermore, the corresponding autonomous system has the origin as a global center. For related results see [42],[110].

Example 4.2. We deal with the forced nonlinear second order scalar equation:

$$x'' + \psi(x) = p(t), \tag{34}$$

where $\psi : \mathbf{R} \to \mathbf{R}$ is continuous and $p : \mathbf{R} \to \mathbf{R}$ is continuous, ω-periodic and such that

$$\bar{p} := (1/\omega) \int_0^\omega p(s)ds = 0.$$

As is well-known, equation (34) is equivalent to the phase-plane system:

$$x' = y + P(t), \qquad y' = -\psi(x),$$

where

$$P(t) := \int_0^t p(s)ds.$$

We assume that the function ψ satisfies:

$$\psi(x) \cdot x > 0 \qquad \text{for} \quad |x| \neq 0, \tag{35}$$

$$\lim_{|x| \to +\infty} \Psi(x) = +\infty, \qquad \text{with} \quad \Psi(x) := \int_0^x \psi(\xi)d\xi\,. \tag{36}$$

From (35) and (36), it follows that the origin in $\mathbf{R}^2$ is a global center for the autonomous system

$$(x', y') = g(x, y) := (y, -\psi(x)) \tag{37}$$

so that there is no compact isolating neighbourhood $\bar{G}$ of the origin (with G open).

Moreover, for any open bounded set $G \subset \mathbf{R}^2$,

$$\deg(g, G, 0) = 1 \quad \text{for} \quad 0 \in G, \qquad \deg(g, G, 0) = 0 \quad \text{for} \quad 0 \notin G.$$

Hence, Theorem 4.1 in [140] cannot be applied.

On the other hand, in order to use Corollary 4.4 it is sufficient to find a bound set G for (37), i.e. an open bounded set with $0 \in G$ such that there is no ω-periodic solution of (37) 'tangent' to ∂G. To this end, we consider, for any $c > 0$, the sublevel set $\Psi_c := \{(x, y) \in \mathbf{R}^2 : (1/2)y^2 + \Psi(x) < c\}$. Then, $\partial \Psi_c$ is a periodic orbit with minimum period:

$$T_c = \sqrt{2} \int_d^c \frac{1}{\sqrt{\Psi(c) - \Psi(\xi)}} d\xi , \qquad \text{with} \quad d < 0 < c,\ \Psi(d) = \Psi(c).$$

Hence, it is sufficient to find $c > 0$ such that $T_c \neq \omega/n$ for all $n \in \mathbf{N}$. Such a choice of c is always possible if the continuous map

$$\tau =: (0, +\infty) \to (0, +\infty), \quad c \mapsto T_c$$

is not constant.

Then, the following result follows from Corollary 4.4 :

Proposition 4.1. *For any continuous map $\psi : \mathbf{R} \to \mathbf{R}$ satisfying* (35), (36) *and having a non-constant associated time-map τ, there is $\varepsilon_0 > 0$ such that equation* (34) *has an ω-periodic solution for every ω-periodic forcing term $p(\cdot)$ with $|p|_1 \leq \varepsilon_0$.*

Recall that, if ψ is continuously differentiable and odd, then, by a classical theorem of Urabe [134] §13.3, Cor. 4, $\tau(\cdot)$ is constant if and only if $\psi : \mathbf{R} \to \mathbf{R}$ is linear.

4.4 Asymptotically homogeneous systems

In this section, we deal with perturbations of autonomous systems with positively homogeneous nonlinearity. More precisely, we consider equations of the form

$$x' = g(x) + e(t, x), \tag{38}$$

with $g : \mathbf{R}^n \to \mathbf{R}^n$ continuous and such that, for some $\alpha > 0$,

(L_1) $\qquad g(kx) = k^\alpha g(x), \quad$ for all $k > 0$, $x \in \mathbf{R}^n$

and $e : [0, \omega] \times \mathbf{R}^n \to \mathbf{R}^n$ satisfying the Caratheodory conditions and such that

(L_2) $\qquad \lim_{|x| \to +\infty} (|e(t, x)|/|x|^\alpha) = 0, \quad$ uniformly a.e. in $t \in [0, \omega]$.

Systems of the form (38) have been widely studied; see, for instance, [64],[69],[71],[87], [102], [103]. In [64]. §41, [102] the more general case in which the function g may depend on t is considered too. However, as we show below, there are situations that can be settled in the framework of Corollary 4.4 but do not fit in [64],[102].

In the first result of this subsection we consider the case of g homogeneous of degree one.

Corollary 4.5. *Assume (L_1), (L_2) with $\alpha = 1$. Suppose that the following conditions are satisfied:*

(L_3) $\qquad$ *$x = 0$ is the only ω-periodic solution of equation*

$$x' = g(x); \tag{39}$$

(L_4) $\qquad \deg(g, B(0, R_0), 0) \neq 0, \qquad$ *for some* $\; R_0 > 0.$
Then there is at least one ω*-periodic solution to* (38).

Note that, from (L_3), the origin is the unique singular point of g, and (L_1), (L_4) imply that $\deg(g, B(0, R), 0) \neq 0$, for every $R > 0$.

Proof. We apply Theorem 4.1, with $G = B(0, R)$, $R > 0$ sufficiently large, $h(t, x; \lambda) = g(x) + \lambda e(t, x)$, and show that for sufficiently large R all the possible solutions of (17)-(16) belong to G. Assume by contradiction that there is a sequence (x_n) of ω-periodic functions, with $|x_n|_\infty \to +\infty$ and such that, for every $n \in \mathbb{N}$,

$$x_n' = g(x_n) + \lambda_n e(t, x_n), \qquad \lambda_n \in [0, 1]. \tag{40}$$

Now we set, for all $n \in \mathbb{N}$,

$$y_n(\cdot) := x_n(\cdot)/|x_n|_\infty ,$$

so that, dividing (40) by $|x_n|_\infty$ and using (L_1) we get

$$y_n' = g(y_n) + (\lambda_n e(t, x_n)/|x_n|_\infty) . \tag{41}$$

We observe that we can apply Ascoli-Arzela's theorem; therefore, there exists y^*, with $|y^*|_\infty = 1$, such that (up to subsequences) $y_n \to y^*$ uniformly on $[0, \omega]$. Thus, taking the limit as $n \to +\infty$ in (41) and using (L_2) we obtain:

$$(y^*)' = g(y^*).$$

Therefore, by (L_3), $y^* = 0$, which is a contradiction.

Thus, we have proved that there is $R > 0$ such that, for every ω-periodic solution x of (17), $x(t) \in B(0, R)$ for all $t \in [0, \omega]$. Therefore, using (L_4) we see that Theorem 4.1 is applicable and we get the existence of an ω-periodic solution of (38) such that $x(t) \in B(0, R)$ for all $t \in [0, \omega]$. ■

Remark 4.3. Corollary 4.5 is a generalization of the results (in the case $\alpha = 1$) in [103], where, besides $(L_1), (L_2)$ and (L_4), the fact that there are no cycles or nontrivial equilibrium states for the autonomous system $x' = g(x)$ is assumed. In other words, (L_3) must hold for periodic solutions of *any* period. Hence, the range of applicability of our corollary is wider than Muhamadiev's one. For instance, Muhamadiev's theorem does not apply to Example 4.1 in Section 4.2; indeed, in that situation, for $\mu, \nu > 0$, the origin is a global center for the autonomous system

$$x_1' = x_2, \qquad x_2' = -\mu x_1^+ + \nu x_1^- ,$$

while our result applies provided that (μ, ν) does not belong to the DANCER-FUČIK spectrum [41].

We also point out that, apparently, Corollary 4.5 (or, more precisely, the evaluation of the Leray-Schauder degree of the Nemitzkii operator induced by (39) in terms of the Brouwer degree of g) cannot be obtained by means of the techniques developed in [64]. To this regard, see the problem raised in [64], p. 256.

Now, we state the analogue of Corollary 4.5 for n-th order systems of the form

$$x^{(m)} + F(x, x', ..., x^{(m-1)}) = e(t, x, x', ..., x^{(m-1)}). \tag{42}$$

Corollary 4.6. *Assume that the following conditions are satisfied :*
(f_1) $F(kx, kx', ..., kx^{(m-1)}) = kF(x, x', ..., x^{(m-1)})$,
for all $k > 0$ *and* $(x, x', ..., x^{(n-1)}) \in \mathbf{R}^{nm}$;
(f_2) $\lim_{|x|+|x'|+...+|x^{(m-1)}| \to +\infty} |e(t, x, x', ..., x^{(m-1)})|/(|x| + |x'| + ... + |x^{(n-1)}|) = 0$,
uniformly a.e. in $t \in [0, \omega]$;
(f_3) $x = 0$ *is the only* ω*-periodic solution of*

$$x^{(m)} + F(x, x', ..., x^{(m-1)}) = 0.$$

(f_4) $\deg(g, B(0, r), 0) \neq 0$, *for* $r > 0$, *where* $g(x) := F(x, 0, ..., 0)$.
Then, (42) has at least one ω*-periodic solution.*

As for the proof, it is sufficient to repeat the argument in the proof of Corollary 4.5 and to apply Corollary 4.3.
Remark 4.4. The particular case when $g(x) = F(x, 0, ..., 0) = \nabla V(x)$, with $V : \mathbf{R}^n \to \mathbf{R}$ a positively homogeneous potential of degree 2 has been considered by many authors (see, e.g., [64], §12.4, [71]). Corollary 4.6 improves the result in [71], where system

$$x'' + \nabla V(x) = p(t)$$

is studied. Indeed. besides assumptions analogous to $(f_1), (f_2), (f3)$, it is assumed in [71] that $V(x) > 0$ for $x \neq 0$ so that (f_4) holds as well (see [71], Th. 12.6). This remark shows that Corollary 4.6 contains the classical theorems in [22],[41] on jumping nonlinearities, where asymptotically homogeneous autonomous equations are considered, and some of the results in [40].[39]. An easier proof of the theorem in [71] has been recently obtained in [93].

As a second consequence of Theorem 4.1, we give a result for asymptotically positively homogeneous systems of order α, with $\alpha \neq 1$.

Corollary 4.7. *Assume* $(L_1), (L_2), (L_4)$ *and suppose, respectively, that either*
(L_3') $x = 0$ *is the only bounded solution of* $x' = g(x)$ *(if* $\alpha > 1$*);*
or
(L_3'') $g(x) \neq 0$ *for* $x \neq 0$ *(if* $\alpha < 1$*).*
Then, (38) has at least one ω*-periodic solution.*

The proof can be obtained by repeating essentially the proof of Corollary 4.5, or following Muhamadiev's argument [103].

By means of [103] and Theorem 4.1 it is easy to extend the result to systems of the form

$$x' = g(t, x) + e(t, x), \tag{43}$$

using an homotopy between (43) and either the 'freezed' system $x' = g(0, x)$ for $\alpha > 1$, or the 'averaged' system $x' = \bar{g}$ for $\alpha < 1$.

Finally, we mention that by means of the Theorem 4.1 it is possible to obtain a result on the so-called *regular guiding functions* (see [7],[64], §14). In this way, we can easily recover [17]. [21].[50], where a perturbation of a polynomial in $\mathbf{R}^2$ is studied.

5 Boundary value problems for some superlinear differential equations

5.1 Introduction

We shall consider here the existence and multiplicity of solutions of nonlinear ordinary differential equations of the form

$$u''(t) + g(u(t)) = p(t, u(t), u'(t)),\ t \in [a, b], \tag{44}$$

satisfying boundary conditions of the Sturm-Liouville or periodic type at a and b, when $g : \mathbf{R} \to \mathbf{R}$ is continuous and *superlinear*, i.e.

$$\frac{g(u)}{u} \to +\infty \text{ as } |u| \to +\infty, \tag{45}$$

and $p : [a, b] \times \mathbf{R}^2 \to \mathbf{R}$ is continuous and satisfies a linear growth condition in the last two arguments. Problems of this type have been considered since the late fifties, among others, by EHRMANN [32], MORRIS [101], FUČIK-LOVICAR [42], STRUWE [133] using shooting arguments and by BAHRI-BERESTYCKI ([2],[3]), RABINOWITZ [120], LONG [78] using critical point theory. The reader can consult [12] for more details and references. It may look surprising that one had to wait for the nineties to see the method of Leray-Schauder applied to such problems (see [12]). The reason can be found in the fact that the success of the Leray-Schauder method relies upon the obtention of a priori estimates for the possible solutions of a family of equations connecting (44) to a simpler problem for which the corresponding topological degree is not zero. A natural choice is of course the homotopy

$$u''(t) + g(u(t)) = \lambda p(t, u(t), u'(t)),\ \lambda \in [0, 1], \tag{46}$$

which reduces to (44) when $\lambda = 1$ and to the simple autonomous equation

$$u''(t) + g(u(t)) = 0, \tag{47}$$

when $\lambda = 0$. An elementary study of (47) under condition (45), based upon the first integral of energy, reveals that (47) will have infinitely many solutions with arbitrary large amplitudes, satisfying the boundary conditions, and the above mentioned methods show indeed, under suitable conditions upon p, that this property holds for all equations (46). Hence, the set of possible solutions of (46) verifying the boundary conditions is not a priori bounded.

A variant of the Leray-Schauder continuation theorem, introduced in [12] to overcome this difficulty in the case of periodic boundary conditions, will be described in the next subsection. Although this approach covers more general situations, and in particular the T-periodic solutions of some perturbed planar Hamiltonian systems of the form

$$z'(t) = -J[H'(z(t)) + p(t, z(t))]$$

with H satisfying some superquadratic condition, we will just describe, in this introduction, the underlying ideas in the special case of equation (44) and for a bounded perturbation p. Let us introduce a continuous function $q : [a,b] \times \mathbf{R}^2 \times [0,1] \to \mathbf{R}$ such that

$$q(t,u,v,1) = p(t,u,v),$$

and such that the set of possible solutions of the periodic problem

$$u''(t) + g(u(t)) = q(t,u(t),u'(t),0),$$

$$u(a) - u(b) = u'(a) - u'(b) = 0,$$

is a priori bounded and the corresponding topological degree of the set of solutions in the space $C^1_\sharp([a,b])$ of C^1 periodic functions on $[a,b]$ is different from zero. This is the case in particular if we choose

$$q(t,u,v,\lambda) = -(1-\lambda)\frac{v}{1+|v|} + \lambda p(t,u,v).$$

Corollaries 2.1 and 2.2 imply then the existence of a continuum $\mathcal{C}$ of solutions (u,λ) of (46) which either connects $C^1_\sharp([a,b]) \times \{0\}$ to $C^1_\sharp([a,b]) \times \{1\}$ (in which case we obtain a periodic solution of (44) or is unbounded in $C^1_\sharp([a,b]) \times [0,1]$. The second possibility is excluded by exhibiting a continuous functional $\varphi : C^1_\sharp([a,b]) \times [0,1] \to \mathbf{R}_+$ which is proper on the set $\Sigma \subset C^1_\sharp([a,b]) \times [0,1]$ of solutions (u,λ) of

$$u''(t) + g(u(t)) = q(t,u(t),u'(t),\lambda), \; u(a) - u(b) = u'(a) - u'(b) = 0, \tag{48}$$

and takes integer values on Σ when $\|u\|$ is sufficiently large. Namely, φ is chosen in such a way that it reduces, for $(u,\lambda) \in \Sigma$ with $\|u\|$ sufficiently large to the winding number around the origin of the curve $\{(u(t),u'(t)) : t \in [a,b]\}$.

In contrast with the techniques mentioned in the first paragraph of this introduction, the present methodology can be and has been extended to the study of periodic solutions of functional differential equations of the form

$$u''(t) + g(u(t)) = p(t,u_t,u'_t),$$

and in particular to delay-differential equations (see [13]). On the other hand, in the Hamiltonian case where p depends only upon t and u, MORRIS, BAHRI-BERESTYCKI, RABINOWITZ and LONG have succeeded in proving the existence of *infinitely many* periodic solutions of (44). One could think of adapting the above continuation technique to obtain such a result by considering instead of (48) the homotopy

$$u''(t) + g(u(t)) = \lambda p(t,u(t)), \; \lambda \in [0,1],$$

$$u(a) - u(b) = u'(a) - u'(b) = 0,$$

and showing the existence of distinct continua of solutions (u,λ) connecting infinitely many distinct periodic solutions of (47) to $C^1_\sharp([a,b]) \times \{1\}$. Unfortunately, the local index in $C^1_\sharp([a,b])$ of any nonconstant periodic solution of (47) is equal to zero (as shown in Section 3 in greater generality), and hence one is unable to prove that such continua merely start from $C^1_\sharp([a,b]) \times \{0\}$!

The situation is quite different in the case of Sturm-Liouville conditions and we shall sketch here for, say. Dirichlet boundary conditions, how to modify the ideas of [12] to prove the existence of infinitely many solutions. We like to mention also that the methodology we have just described has been successfully applied by FURI and PERA [47] to study the periodic solutions of the forced spherical pendulum and that a general abstract version has been given by CAPIETTO in her Ph.D. thesis [9].

5.2 A continuation theorem for semilinear equations in Banach spaces

In this section, we shall introduce a generalization of the continuation theorem of [12]. Let X, Z be real Banach spaces, $L : D(L) \subset X \to Z$ a linear Fredholm mapping of index zero, $I = [0,1]$ and $N : X \times I \to Z$ a L-completely continuous operator. We consider the equation

$$Lu = N(u,\lambda),\ u \in D(L),\ \lambda \in I. \tag{49}$$

For any set $B \subset X \times I$ and any $\lambda \in I$ we again denote by B_λ the section $\{u \in X : (u,\lambda) \in B\}$. Let $\mathcal{O} \subset X \times I$ be open in $X \times I$, with $\overline{\mathcal{O}}$ and $\partial\mathcal{O}$ its closure and boundary in $X \times I$ respectively. Let us denote by Σ the (possibly empty) set of solutions (u,λ) of (6) which belong to $\overline{\mathcal{O}}$, i.e.

$$\Sigma = \{(u,\lambda) \in \overline{\mathcal{O}} \cap (D(L) \times I) : Lu = N(u,\lambda)\}.$$

We first suppose that

(i_1) *Σ_0 is bounded in X and $\Sigma_0 \subset \mathcal{O}_0$,*

so that, for any open bounded set $\Omega \subset X$ with $\Sigma_0 \subset \Omega \subset \mathcal{O}_0$, the degree $D_L(L-N(.,0),\Omega)$ is well defined and independent of Ω. We further assume that

(i_2) $D_L(L - N(.,0).\Omega) \neq 0,$

so that $\Sigma_0 \neq \emptyset$. Finally, we introduce a functional $\varphi : X \times I \to \mathbf{R}$ and suppose that

(i_3) *φ is continuous on $X \times I$ and proper on Σ.*

Consequently, the constants

$$\varphi_- = \min\{\varphi(u,0) : u \in \Sigma_0\},\ \varphi_+ = \max\{\varphi(u,0) : u \in \Sigma_0\}$$

exist.

Theorem 5.1. *Assume that conditions (i_1), (i_2) (i_3) hold and that there exist constants c_-, c_+ with*

$$c_- < \varphi_- \leq \varphi_+ < c_+,$$

such that

$$\varphi(u,\lambda) \notin \{c_-, c_+\},$$

whenever $(u,\lambda) \in (D(L)\times]0,1[) \cap \mathcal{O} \cap \Sigma$, *and*

$$\varphi(u,\lambda) \notin [c_-, c_+],$$

whenever $(u,\lambda) \in (D(L)\times]0,1[) \cap \partial\mathcal{O} \cap \Sigma$.
Then equation

$$Lu = N(u,1) \tag{50}$$

has at least one solution in $D(L) \cap (\overline{\mathcal{O}})_1$.

Proof. Assume that equation (50) has no solution. Then

$$\varphi(\mathcal{O} \cap \Sigma) \cap \{c_-, c_+\} = \emptyset, \tag{51}$$

and

$$\varphi(\partial\mathcal{O} \cap \Sigma) \cap [c_-, c_+] = \emptyset. \tag{52}$$

Corollary 2.2 asserts that there exists a continuum $\mathcal{C} \subset \Sigma$ with $\mathcal{C} \cap (\Sigma_0 \times \{0\}) \neq \emptyset$ such that either $\mathcal{C}$ is unbounded or $\mathcal{C}$ intersects $\partial\mathcal{O}$. By our assumption, $\varphi(\mathcal{C})$ is connected and intersects $[\varphi_-, \varphi_+]$ as $\mathcal{C} \cap (\Sigma_0 \times \{0\}) \neq \emptyset$. Suppose that $\mathcal{C}$ intersects $\partial\mathcal{O}$. Then the interval $\varphi(\mathcal{C})$ intersects at least one of the intervals $]-\infty, c_-[$ or $]c_+, +\infty[$. Hence $\{c_-, c_+\} \cap \varphi(\mathcal{C}) \neq \emptyset$ and if $(\tilde{u}, \tilde{\lambda}) \in \mathcal{C}$ is such that $\varphi(\tilde{u}, \tilde{\lambda}) = c \in \{c_-, c_+\}$, then $(\tilde{u}, \tilde{\lambda}) \in \bar{\mathcal{O}}$ but this is excluded by (51) and (52). Suppose now that $\mathcal{C}$ is unbounded, so that $\varphi(\mathcal{C})$ is unbounded too. then, at least one of the unbounded intervals $]-\infty, \varphi_-]$, $[\varphi_+, +\infty[$ is contained in $\varphi(\mathcal{C}) \subset \varphi(\Sigma)$, which implies that $\varphi(\Sigma) \cap \{c_-, c_+\} \neq \emptyset$, a contradiction. ■

Remark 5.1. If $\Omega \subset X$ is open, bounded and such that

$$Lx \neq N(x,\lambda) \textit{ for all } (x,\lambda) \in \partial\Omega \times [0,1[,$$

and if we take $\mathcal{O} = \Omega \times [0,1]$ and

$$\begin{aligned} \varphi(x,\lambda) &= -\mathrm{dist}(x, \partial\Omega) \text{ if } x \in \Omega, \\ &= \mathrm{dist}(x, \partial\Omega) \text{ if } x \notin \Omega, \end{aligned}$$

then, by assumption, $\Sigma_0 \subset \Omega$ is bounded, $-\mathrm{diam}\,\Omega \leq \varphi_+ < 0$, and we can take $c_+ = 0, c_- < -\mathrm{diam}\,\Omega$ in the above theorem to recover the classical Leray-Schauder continuation principle.

Let us now consider a consequence of Theorem 5.1 which is useful for the application we have in mind. Let

$$\Sigma^* = \{(u,\lambda) \in D(L) \times I : Lu = N(u,\lambda)\}.$$

Assume that $\varphi : X \times I \to \mathbf{R}_+$ is continuous and satisfies the following conditions

(i_4) *There exists* $R > 0$ *such that* $\varphi(u, \lambda) \in \mathbf{N}$ *for all* $(u, \lambda) \in \Sigma^*$ *with* $\|u\| \geq R$.

(i_5) $\varphi^{-1}(n) \cap \Sigma^*$ *is bounded for each* $n \in \mathbf{N}$.

Let k_0 be an integer such that

$$k_0 > \sup\{\varphi(u, \lambda) : (u, \lambda) \in \Sigma^*, \|u\| \leq R\}.$$

Then $\Phi_0 = \varphi^{-1}([0. k_0])$ is a closed subset of $X \times I$ containing $B[R] \times I$. For each integer $k > k_0$, $\Sigma^k = \varphi^{-1}(k) \cap \Sigma^*$ is a compact set such that $\Sigma^k \cap \Phi_0 = \emptyset$ and hence $\mathrm{dist}(\Sigma^k, \Phi_0) > 0$. Choose, for those k,

$$\epsilon_k \in]0, \min\{1/2, (1/2)\mathrm{dist}(\Sigma^k, \Phi_0)\}]$$

and set

$$\mathcal{O}^k = \{(u, \lambda) \in X \times I : |\varphi(u, \lambda) - k| < \epsilon_k\}.$$

Fix an integer $j > k_0$ and consider the open set

$$\mathcal{O}_j = (X \times I) \setminus (\Phi_0 \cup \bigcup_{k > k_0, k \neq j} \overline{\mathcal{O}^k}).$$

Then,

$$\Sigma^j = \{(u, \lambda) \in \overline{\mathcal{O}_j} \cap (D(L) \times I) : Lu = N(u, \lambda)\},$$

and, by construction, $\varphi(\Sigma^j \cap \mathcal{O}_j) = j$, so that

$$\varphi_-^j := \min\{\varphi(u, 0) : u \in \Sigma_0^j\} = \varphi_+^j := \max\{\varphi(u, 0) : u \in \Sigma_0^j\} = j.$$

Finally, $\Sigma_0^j = (\varphi^{-1}(j))_0 \cap \Sigma_0^*$ is bounded and $\Sigma_0^j \subset (\mathcal{O}_j)_0$. Let us assume that

(i_6) $D_L(L - N(., 0), \mathcal{V}_j) \neq 0$ *for some bounded open* $\mathcal{V}_j \supset \Sigma_0^j$.

Then, if we take $c_-^j = j - \epsilon_j$, $c_+^j = j + \epsilon_j$, we have $\varphi(u, \lambda) \notin \{c_-^j, c_+^j\}$ for each $(u, \lambda) \in (D(L) \times]0, 1[) \cap \mathcal{O}_j \cap \Sigma^j$ and $\varphi(u, \lambda) \notin [c_-^j, c_+^j]$ for each $(u, \lambda) \in (D(L) \times]0, 1[) \cap \partial\mathcal{O}_j \cap \Sigma^j$. Thus, all conditions of Theorem 1 with $\Sigma = \Sigma^j$, $\mathcal{O} = \mathcal{O}_j$ are satisfied and equation (50) will have at least one solution $u \in D(L) \cap (\overline{\mathcal{O}^j})_1$. We have therefore the following result.

Corollary 5.1. *Assume that conditions* (i_4) *and* (i_5) *hold and that* (i_6) *is satisfied for each integer* $j > k_0$. *Then, for each of those integers, equation (50) has at least one solution* u_j *such that* $\varphi(u_j, 1) = j$. *Moreover,* $\lim_{j \to \infty} \|u_j\| = +\infty$.

Proof. Only the last part of the Corollary has still to be proved. If this conclusion is not true, we can find a bounded subsequence (u_{j_k}) of solutions of (50) with $\varphi(u_{j_k}) = j_k \to \infty$ as $k \to \infty$. Thus we get a contradiction, as the sequence (u_{j_k}) is precompact.

5.3 A candidate for the functional φ for periodic problems and its properties

Let $h : \mathbf{R} \times \mathbf{R}^2 \times I \to \mathbf{R}^2, (t, x, \lambda) \mapsto h(t, x, \lambda)$ be T-periodic in t and continuous. Let

$$J = \begin{pmatrix} 0 & -1 \\ 1 & 0 \end{pmatrix}$$

be the symplectic matrix in $\mathbf{R}^2$ and let

$$C_T = \{z \in C(\mathbf{R}, \mathbf{R}^2) : z(t+T) = z(t) \text{ for all } t \in \mathbf{R}\}$$

with the norm $\|z\| = \max_{t \in \mathbf{R}} |z(t)|$, $|z(t)|$ denoting the Euclidian norm of $z(t)$ in $\mathbf{R}^2$. Define $\delta : \mathbf{R}^2 \to \mathbf{R}$ by $\delta(x) = 1$ if $|x| < 1$ and $\delta(x) = |x|^{-2}$ if $|x| \geq 1$, and $\varphi : C_T \times [0,1] \to \mathbf{R}_+$ by

$$\varphi(z, \lambda) = \frac{1}{2\pi} \left| \int_0^{2\pi} (h(t, z(t), \lambda), Jz(t)) \delta(z(t))\, dt \right| \tag{53}$$

where $(.,.)$ denotes the inner product in $\mathbf{R}^2$. Let

$$D(L) = C_T \cap C^1(\mathbf{R}, \mathbf{R}^2), L : D(L) \subset C_T \to C_T, z \mapsto z',$$
$$N : C_T \times [0,1] \to C_T, z \mapsto h(., z(.), \lambda),$$
$$\Sigma = \{(z, \lambda) \in D(L) \times [0,1] : Lz = N(z, \lambda)\}.$$

It is routine to check that φ is continuous and N is L-completely continuous on $C_T \times [0,1]$ (see [89]).

Proposition 5.1. *If $(z, \lambda) \in \Sigma$ and $\min_{t \in \mathbf{R}} |z(t)| \geq 1$, then $\phi(z, \lambda) \in \mathbf{Z}_+$.*

Proof. Letting $z_1(t) = \rho(t) \cos \theta(t)$, $z_2(t) = \rho(t) \sin \theta(t)$, we get

$$\begin{aligned} \theta'(t) &= [z_2'(t) z_1(t) - z_1'(t) z_2(t)] |z(t)|^{-2} \\ &= (h(t, z(t), \lambda), Jz(t)) |z(t)|^{-2} \\ &= (h(t, z(t), \lambda), Jz(t)) \delta(z(t)), \end{aligned} \tag{54}$$

and hence

$$\phi(z, \lambda) = \frac{1}{2\pi} \left| \int_0^T \theta'(t)\, dt \right| = \frac{|\theta(T) - \theta(0)|}{2\pi} \in \mathbf{Z}_+.$$

■

Proposition 5.2. *If there exists $R > 0$ such that*

$$(h(t, x, \lambda), Jx) \geq S(x) - \gamma |x|, \tag{55}$$

or

$$(h(t, x, \lambda), Jx) \leq -S(x) + \gamma |x|, \tag{56}$$

for some $\gamma \geq 0$, some positive definite quadratic form S on $\mathbf{R}^2$, all $t \in \mathbf{R}$, all $\xi \in C$, all $\lambda \in [0,1]$ and all $x \in \mathbf{R}^2$ such that $|x| \geq R$, then there exists $R_1 \geq 1$ such that

$$\varphi(z, \lambda) \geq (\omega \langle S \rangle)^{-1},$$

for all $(z,\lambda) \in \Sigma$ *with* $\min_{t\in\mathbf{R}} |z(t)| \geq R_1$, *where* $\omega = 2\pi/T$ *and*

$$\langle S \rangle = \frac{1}{2\pi} \int_0^{2\pi} \frac{d\theta}{S(\cos\theta, \sin\theta)}.$$

Proof. Assume for definiteness that (55) holds. Let $A \geq \max\{1, R\}$ and $\sigma = \min_{S^1} S$ and assume that $(z,\lambda) \in \Sigma$ and $|z(t)| \geq A$ for all $t \in \mathbf{R}$. Then, using (54) and (55), we get

$$\theta'(t) \geq S(\cos\theta(t), \sin\theta(t)) - \gamma/A,$$

and hence

$$\int_0^T \frac{\theta'(t)}{S(\cos\theta(t), \sin\theta(t))}\, dt \geq T - \frac{\gamma T}{\sigma A},$$

i.e.

$$\int_{\theta(0)}^{\theta(T)} \frac{d\theta}{S(\cos\theta, \sin\theta)} \geq T - \frac{\gamma T}{\sigma A}. \tag{57}$$

Now, $\theta(T) - \theta(0) = 2k\pi$ for some $k \in \mathbf{Z}$ with

$$\varphi(x,\lambda) = |k|.$$

Hence, (57) becomes

$$k \geq (\omega\langle S \rangle)^{-1} - \frac{\gamma}{\omega\sigma A\langle S \rangle}. \tag{58}$$

Thus, if we take $A \geq \gamma/\sigma$, k is a nonnegative integer, so that

$$\varphi(z,\lambda) = k, \tag{59}$$

and, moreover, taking still A larger if necessary and using the fact that k is an integer, (58) and (59) imply that

$$\varphi(z,\lambda) \geq (\omega\langle S \rangle)^{-1}.$$

■

Proposition 5.3. *If there exists* $V \in C^1(\mathbf{R}^2, \mathbf{R})$ *and* $r > 0$ *such that*
1. $|V(x)| \to +\infty$ as $|x| \to \infty$.
2. $(V'(x), h(t,x,\lambda)) \leq c|V(x)|$ *for some* $c > 0$ *and all* x *with* $|x| \geq r$.
Then, for each $r_1 > 0$, *there exists* $r_2 \geq r_1$ *such that for each* $(z,\lambda) \in \Sigma$ *for which* $\|z\| > r_2$, *one has* $\min_{t\in\mathbf{R}} |z(t)| > r_1$.

Proof. By assumption 1, we can find $r_0 > r$ such that $|V(x)| > 0$ whenever $|x| \geq r_0$. Define W for those x by

$$W(x) = \log|W(x)|$$

and observe that

$$\lim_{|x|\to\infty} W(x) = +\infty \tag{60}$$

and

$$W'(x) = (V(x))^{-1}V'(x),$$

with $W : \mathbf{R}^2 \setminus B[0, r_0] \to \mathbf{R}$ of class C^1. Let $(z, \lambda) \in \Sigma$ be such that

$$\min_{t \in \mathbf{R}} |z(t)| \leq r_1 \tag{61}$$

and fix $c_1 > \max\{r_0, r_1\}$. Choose t_0 such that $|z(t_0)| = \|z\|$. If $\|z\| > c_1$, then, using (61) and the choice of c_1 we can find $t_1 \in \mathbf{R}$ such that $|z(t_1)| = c_1$ and, by T-periodicity, we can always choose t_0 and t_1, with $|t_0 - t_1| \leq T$, such that $t_0 - t_1$ and $V(x)$ for $|x| \geq r_0$ have the same sign and $|z(t_0)| > c_1$ for $t \in [t_1, t_0]$ or $[t_0, t_1]$. Now, let

$$w(t) = W(z(t)).$$

We have

$$\begin{aligned} w(t_0) &= w(t_1) + \int_{t_1}^{t_0} w'(s)\, ds \leq w(t_1) + c|t_0 - t_1| \\ &\leq \max\{W(x) : |x| = c_1\} + cT = c_2. \end{aligned} \tag{62}$$

From (60), we can find $r_3 \geq r_0$ such that $W(x) > c_2$ for $|x| > r_3$, so that (62) implies that

$$\|z\| = |z(t_0)| \leq r_3,$$

and the result is proved for $r_2 = \max\{c_1, r_3\}$. ■

5.4 Perturbations of planar autonomous Hamiltonian systems and applications

Let $H \in C^1(\mathbf{R}^2, \mathbf{R})$ and $p : \mathbf{R} \times \mathbf{R}^2 \to \mathbf{R}^2, (t, x) \mapsto p(t, x)$ be T-periodic in t, continuous and takes bounded sets into bounded sets.

Theorem 5.2. *Assume that the following conditions hold.*
1. $|H(x)| \to \infty$ as $|x| \to \infty$.
2. There exists $r_0 > 0$ *such that* $H'(x) \neq 0$ *for* $|x| \geq r_0$.
3. There exist $r > 0$ *and* $c > 0$ *such that*

$$(-Jp(t, x), H'(x)) \leq c|H(x)|$$

for all $t \in \mathbf{R}$ *and* $x \in \mathbf{R}^2$ *such that* $|x| \geq r$.
4. For each $\alpha > 0$, *there exists* $d(\alpha) > 0$ *and a positive definite quadratic form* S_α *on* $\mathbf{R}^2$ *such that*

$$\lim_{\alpha \to \infty} \langle S_\alpha \rangle = 0$$

and

$$(H'(x), x) - |(p(t, x), x)| \geq S_\alpha(x) - d(\alpha)$$

for all $t \in \mathbf{R}$ *and* $x \in \mathbf{R}^2$.
Then the system

$$z'(t) = -J[H'(z(t)) + p(t, z(t))] \tag{63}$$

has at least one T-periodic solution.

Proof. Define $h : \mathbf{R} \times \mathbf{R}^2 \times [0,1] \to \mathbf{R}^2$ by

$$h(t,x,\lambda) = -JH'(x) - \lambda Jp(t,x) - (1-\lambda)\frac{H'(x)}{1+|H'(x)|},$$

and L and N correspondingly as in Section 5.3. We shall show that the conditions of Corollary 5.1 are satisfied.

First we have. using assumption 2,

$$(h(t,x,0), H'(x)) = -\frac{|H'(x)|^2}{1+|H'(x)|} < 0$$

for $|x| \geq r_0$, so that H is a strict guiding function (see e.g. [89] for the terminology and properties) for the equation

$$z'(t) = h(t, z(t), 0),$$

which implies by assumption 1 that Σ_0 is bounded and

$$|D_L(L - N(.,0), \Omega)| = 1$$

for any open bounded set $\Omega \subset C_T$ such that $\Omega \supset \Sigma_0$.

Now, using assumption 3, we get

$$(h(t,x,\lambda), H'(x)) = -\lambda(Jp(t,x), H'(x)) - (1-\lambda)\frac{|H'(x)|^2}{1+|H'(x)|} \leq c|H(x)|$$

whenever $|x| \geq r$. Thus, using Proposition 5.2 with $V = H$, and Proposition 5.1, we easily find $\rho > 0$ such that $\phi(z,\lambda) \in \mathbf{Z}_+$ if $(z,\lambda) \in \Sigma \setminus (B(0,\rho) \times [0,1])$.

Furthermore, using assumption 4, we get

$$\begin{aligned}(h(t,x,\lambda). Jx) &= -(JH'(x), Jx) - \lambda(Jp(t,x), Jx) - (1-\lambda)\frac{(H'(x), Jx)}{1+|H'(x)|} \\ &= -(H'(x), x) + |p(t,x), x)| + |x| \\ &\leq -S_\alpha(x) + d(\alpha) + |x| \\ &\leq -S_\alpha(x) + 2|x|,\end{aligned}$$

for all $t \in \mathbf{R}, \xi \in \mathcal{C}$ and $|x| \geq d(\alpha)$. Consequently, for each $\alpha > 0$, there exists, by Proposition 5.2, some $r(\alpha) > 0$ such that

$$\varphi(z,\lambda) \geq (\omega\langle S_\alpha\rangle)^{-1} \tag{64}$$

for all $(z,\lambda) \in \Sigma$ with $\min_{t\in\mathbf{R}} |z(t)| \geq r(\alpha)$. Using Proposition 5.3 again, this implies that for each $\alpha > 0$, there exists $R(\alpha) > 0$ such that (64) holds whenever $(z,\lambda) \in \Sigma$ and $\|z\| \geq R(\alpha)$. As

$$(\langle S_\alpha\rangle)^{-1} \to +\infty \; if \; \alpha \to +\infty,$$

we easily obtain that $\phi^{-1}(n) \cap \Sigma$ is bounded for each $n \in \mathbf{Z}_+$, and the result follows from Corollary 5.1.

■

As a first application, let us consider the second order differential equation

$$x''(t) + g(x(t)) = e(t, x(t), x'(t)), \tag{65}$$

where $g : \mathbf{R} \to \mathbf{R}$ and $e : \mathbf{R} \times \mathbf{R}^2 \to \mathbf{R}$ are continuous and e is T-periodic in t.

Theorem 5.3. *Assume that the following conditions hold.*
1.

$$\lim_{|u|\to\infty} \frac{g(u)}{u} = +\infty. \tag{66}$$

2. There exists $K \geq 0, L \geq 0$ such that

$$|e(t, u, v)| \leq K(|u| + |v|) + L$$

for all $(t, u, v) \in \mathbf{R} \times \mathbf{R}^2$.
Then the equation (65) has at least one T-periodic solution.

Proof. Letting $z_1(t) = u(t), z_2(t) = u'(t)$, we can write (65) in the form (63) with

$$H(x) = H(x_1, x_2) = G(x_1) + (1/2)x_2^2, G(u) = \int_0^u g(s)\, ds,$$

and

$$p(t, x) = p(t, x_1, x_2) = (-e(t, x_1, x_2), 0).$$

Clearly, $H(x) \to +\infty$ as $|x| \to \infty$ and $H'(x) \neq 0$ for $|x|$ large, by (66) and its consequence $\lim_{|u|\to\infty} u^{-2}G(u) = +\infty$. Indeed, $G(u) \geq (u^2/2) - c_0$ for some $c_0 \geq 0$, and $H(x) \geq (1/2)|x|^2 - c_0$ for all $x \in \mathbf{R}^2$. Hence, for $|x| \geq 2\sqrt{c_0}$, we have

$$\frac{-(Jp(t,x), H'(x))}{H(x)} = \frac{e(t,x)x_2}{H(x)} \leq \frac{(K(|x_1| + |x_2|) + L)|x_2|}{(1/2)|x|^2 - c_0} \leq c,$$

for some $c > 0$.

On the other hand, for each $\alpha > 0$, there exists $\delta(\alpha) > 0$ such that

$$g(u)u \geq (\alpha^2 + \frac{K^2 + 1}{2} + K)u^2 - \delta(\alpha),$$

for all $u \in \mathbf{R}$ and hence

$$\begin{aligned}
(H'(x), x) - |(p(t,x), x)| &\geq g(x_1)x_1 + x_2^2 - [K(|x_1| + |x_2|) + L]|x_1| \\
&\geq (\alpha^2 + \frac{K^2+1}{2})x_1^2 \\
&- K|x_1||x_2| - L|x_1| - \delta(\alpha) \\
&\geq \alpha^2 x_1^2 + \frac{x_2^2}{2} - \delta(\alpha) - \frac{L^2}{2} \\
&= \alpha^2 x_1^2 + \frac{x_2^2}{2} - d(\alpha),
\end{aligned}$$

and we can take $S_\alpha = \alpha^2 x_1^2 + \frac{x_2^2}{2}$ for which $\langle S_\alpha \rangle = \frac{\sqrt{2}}{\alpha} \to 0$ as $\alpha \to +\infty$. Thus all conditions of Theorem 5.2 hold and the result follows. ∎

As another application we consider the perturbed Hamiltonian system

$$z'(t) = -J[H'(z(t)) + q(t, z(t))], \tag{67}$$

where $H \in C^1(\mathbf{R}^2, \mathbf{R})$ and $q : \mathbf{R} \times \mathbf{R}^2 \to \mathbf{R}^2$ is T-periodic in t and continuous.

Theorem 5.4. *Assume that the following conditions hold.*
1. $(H'(x), x)/|x|^2 \to +\infty$ as $|x| \to \infty$.
2. $|H'(x)| \leq A|H(x)| + B$ for some $A > 0, B \geq 0$ and all $x \in \mathbf{R}^2$.
3. $|q(t, x)| \leq C$ for some $C > 0$ and all $(t, x) \in \mathbf{R} \times \mathbf{R}^2$.
Then the system (67) has at least one T-periodic solution.

Proof. Condition 1 implies that, for $x \neq 0$, one has

$$|H'(x)| \geq (H'(x), x)/|x| \to +\infty \; as \; |x| \to \infty,$$

and hence the same is true for $H(x)$ by condition 2, so that assumptions 1 and 2 of Theorem 5.2 hold. Now,

$$(-Jq(t, x), H'(x)) \leq C|H'(x)| \leq AC|H(x)| + BC \leq (AC + 1)|H(x)|,$$

if $|x|$ is large enough so that $|H(x)| \geq BC$. Finally, if $\delta(\alpha)$ is such that

$$(H(x), x)/|x|^2 \geq (\alpha^2 + \frac{1}{2})|x|^2 - \delta(\alpha),$$

we have

$$\begin{aligned} (H'(x), x) - |(q(t, \xi), x)| &\geq (H'(x), x) - C|x| \\ &\geq (\alpha^2 + \frac{1}{2})|x|^2 - \frac{|x|^2 + C^2}{2} - \delta(\alpha) \\ &= \alpha^2|x|^2 - d(\alpha), \end{aligned}$$

and assumption 4 of Theorem 5.2 is easily checked. ∎

5.5 Superlinear second order equations with Dirichlet conditions

We want now to apply the abstract theory to the problem

$$u''(t) + g(u(t)) = p(t, u(t), u'(t)), \; u(0) = u(1) = 0,$$

when (45) holds and $p : [0, 1] \times \mathbf{R}^2 \to \mathbf{R}$ is continuous and has a linear growth with respect to the last two variables, Without loss of generality we choose $a = 0$, $b = 1$ and assume that $g(u)u > 0$ for $u \neq 0$ and we consider the homotopy

$$u''(t) + g(u(t)) = \lambda p(t, u(t), u'(t)), \; u(0) = u(1) = 0, \lambda \in I, \tag{68}$$

which can easily be written as an abstract equation of type (49) in the space $C_0^1[0,1] = \{u \in C^1[0,1] : u(0) = u(1) = 0\}$ with the usual C^1 norm by setting $Lu = -u''$, $N(u,\lambda) = g(u(.)) - \lambda p(.,u(.),u'(.))$. We first study the problem when $\lambda = 0$, i.e.

$$u''(t) + g(u(t)) = 0,\ u(0) = u(1) = 0. \tag{69}$$

If $G(u) = \int_0^u g(s)\,ds$, the energy integral shows that all the orbits of the equation

$$u'' + g(u) = 0,$$

in the phase plane (u,u') are closed and surround the origin, and, for each $\alpha \neq 0$, the equation $G(c) = \frac{\alpha^2}{2}$ has two solutions $c_-(\alpha) < 0 < c_+(\alpha)$. Notice that $c_\pm(-\alpha) = c_\pm(\alpha)$. We define moreover the time-maps by

$$\tau_-(\alpha) = 2\int_{c_-(\alpha)}^0 \frac{ds}{\sqrt{\alpha^2 - 2G(s)}},\ \tau_+(\alpha) = 2\int_0^{c_+(\alpha)} \frac{ds}{\sqrt{\alpha^2 - 2G(s)}},$$

so that $\tau_\pm(-\alpha) = \tau_\pm(\alpha)$ and $\tau_\pm(\alpha) \to 0$ as $\alpha \to \infty$. The solution $u(.,\alpha)$ of the Cauchy problem

$$u''(t) + g(u(t)) = 0,\ u(0) = 0, u'(0) = \alpha,$$

will be a solution of (69) if and only if

$$m\tau_-(\alpha) + n\tau_+(\alpha) = 1$$

for some nonnegative integers m, n such that $|m - n| \leq 1$. Let

$$\mathcal{S} = \{(x,y) \in \mathbf{R}^2 : x > 0, y > 0, mx + ny = 1$$

$$\text{for some } (m,n) \in \mathbf{N}^2 \text{ with } |m-n| \leq 1\}.$$

Then $u(.,\alpha)$ is a solution of (69) if and only if $(\tau_-(\alpha), \tau_+(\alpha)) \in \mathcal{S}$. If we define the open bounded set G^α in $\mathbf{R}^2$ by

$$G^\alpha = \{(u,v) \in \mathbf{R}^2 : v^2 + 2G(u) < \alpha^2\}$$

and the corresponding open bounded set Ω^α in $C_0^1[0,1]$ by

$$\begin{aligned}\Omega^\alpha &= \{u \in C_0^1[0,1] : u'^2(t) + 2G(u(t)) < \alpha^2 \text{ for all } t \in [0,1]\} \\ &= \{u \in C_0^1[0,1] : (u(t), u'(t)) \in G^\alpha \text{ for all } t \in [0,1]\},\end{aligned}$$

then equation (69) has no solution on $\partial\Omega^\alpha$ for each α such that $(\tau_-(\alpha), \tau_+(\alpha)) \notin \mathcal{S}$. and hence

$$D_L(L - N(.,0), \Omega^\alpha) = \deg(I - L^{-1}N(.,0), \Omega^\alpha, 0)$$

is well defined. Using a duality theorem of KRASNOSEL'SKII (see [64]), one can prove the following result.

Lemma 5.1. *If α is such that $(\tau_-(\alpha), \tau_+(\alpha)) \notin \mathcal{S}$, then*

$$D_L(L - N(.,0), \Omega^\alpha) = \deg(U,]-\alpha, \alpha[, 0),$$

where $U : \mathbf{R} \to \mathbf{R}$ is defined by $U(c) = u(1; c)$.

The mapping $\alpha \mapsto \deg(U,]-\alpha, \alpha[, 0)$ is constant when $(\tau_-(\alpha), \tau_+(\alpha))$ remains in a connected component of the set $\mathbf{R}^{+2} \setminus \mathcal{S}$ and the corresponding value can be computed.

Lemma 5.2. *For α such that $(\tau_-(\alpha), \tau_+(\alpha))$ belongs to the unbounded component containing the diagonal, $\deg(U,]-\alpha, \alpha[, 0) = 1$ and its value on the bounded components containing the diagonal will then be alternatively minus one and plus one as one approachs the origin. On all the other components, $\deg(U,]-\alpha, \alpha[, 0) = 0$.*

The reader is advised to draw the corresponding picture.

Lemma 5.3. *If $0 < \alpha < \beta$ are such that $(\tau_-(\alpha), \tau_+(\alpha))$ and $(\tau_-(\beta), \tau_+(\beta))$ are not in $\mathcal{S}$, then*

$$D_L(L - N(.,0), \Omega^\beta_\alpha) = D_L(L - N(.,0), \Omega^\beta) - D_L(L - N(.,0), \Omega^\alpha),$$

where

$$\Omega^\beta_\alpha = \{u \in C^1_0[0,1] : (u(t), u'(t)) \in G^\beta \setminus \overline{G^\alpha} \text{ for all } t \in [0,1]\}.$$

We refer to [14] for the proofs of those three results. If we suppose now, for simplicity, that g *is odd*, so that $c_-(\alpha) = c_+(\alpha)$ and $\tau_-(\alpha) = \tau_+(\alpha)$, we find immediately from Lemma 5.1 and Lemma 5.2 that

$$D_L(L - N(.,0), \Omega^\alpha) = (-1)^n,$$

if n is such that $\frac{1}{n+1} < \tau_-(\alpha) < \frac{1}{n}$.

Let

$$\delta : \mathbf{R}^2 \to \mathbf{R}, (u,v) \mapsto \min\{1, \frac{1}{u^2 + v^2}\}.$$

We now define the continuous functional φ on $C^1_0[0,1] \times I$ by

$$\varphi(u,\lambda) = |\frac{1}{\pi} \int_0^1 [u'^2(t) + u(t)g(u(t)) - \lambda p(t, u(t), u'(t))u(t)]\delta(u(t), u'(t))\, dt|.$$

Consequently, if (u, λ) is a solution of (68) such that $u^2(t) + u'^2(t) \geq 1$ for all $t \in [0,1]$, we get, letting $v(t) = u'(t)$,

$$\varphi(u,\lambda) = |\frac{1}{\pi} \int_0^1 \frac{v(t)u'(t) - u(t)v'(t)}{u^2(v) + v^2(t)}\, dt| = |\frac{1}{\pi} \int_0^1 d \operatorname{arctg} \frac{v(t)}{u(t)}|.$$

Thus $\varphi(u,\lambda) + 1$ is equal to the number of zeroes of u on $[0,1]$ when the integrand is positive.

On the other hand, the following properties of φ can be proved like in [12]. Consistently with the notations of Corollary 5.1, we set

$$\Sigma^* = \{(u,\lambda) \in C^1_0([0,1]) : (u,\lambda) \text{ is a solution of } (10)\}.$$

Proposition 5.4. *There exists $C > 0$ such that for each $\alpha > 0$ one can find $R_1(\alpha) \geq 1$ such that $\varphi(u,\lambda) \geq C\alpha$ whenever $(u,\lambda) \in \Sigma^*$ satisfies the condition $u^2(t) + u'^2(t) \geq R_1^2(\alpha)$ for all $t \in [0,1]$.*

Proposition 5.5. *For each $r_1 > 0$ one can find $r_2 \geq r_1$ such that $u^2(t) + u'^2(t) \geq r_1^2$ for all $t \in [0,1]$ whenever $(u, \lambda) \in \Sigma^*$ is such that $\|u\| \geq r_2$.*

The following consequences are immediate.

Corollary 5.2. *There exists $R > 0$ such that $\varphi(u, \lambda) \in \mathbf{N}$ whenever $(u, \lambda) \in \Sigma^*$ and $\|u\| \geq R$.*

Corollary 5.3. *$\varphi(u, \lambda) \to +\infty$ whenever $(u, \lambda) \in \Sigma^*$ and $\|u\| \to \infty$.*

As a consequence, $\varphi^{-1}(k) \cap \Sigma^*$ is bounded for each $k \in \mathbf{N}$. The above mentioned degree computations can be used to prove that if Ω_k is an open bounded set containing $\varphi^{-1}(k) \cap \Sigma^* \cap X \times \{0\}$, and no other part of Σ_0^*, one has

$$D_L(L - N(.,0), \Omega_k) = 2(-1)^k.$$

The conditions of Corollary 5.1 are satisfied and we have proved the following result.

Theorem 5.5. *Assume that $g : \mathbf{R} \to \mathbf{R}$ is continuous, odd and superlinear and that $p : [0,1] \times \mathbf{R}^2 \to \mathbf{R}$ is continuous and has at most linear growth in the last two variables. Then there exists $k_0 \in \mathbf{N}$ such that, for each $j > k_0$ the problem*

$$u''(t) + g(u(t)) = p(t, u(t), u'(t)), \; u(0) = u(1) = 0$$

has at least one solution u_j having exactly $j+1$ zeros on $[0,1]$. Moreover, $\|u_j\| \to \infty$ as $j \to \infty$.

Remarks. 1. Similar results hold for the Neumann boundary value problem

$$u''(t) + g(u(t)) = p(t, u(t), u'(t)), \; u'(0) = u'(1) = 0,$$

without the oddness condition for g.

2. When g and/or p satisfy suitable symmetry conditions, those results can be used to prove, for equation (44), the existence of infinitely many periodic solutions having related symmetries.

6 Periodic solutions of second order differential equations with singular nonlinearities

6.1 Introduction

Second order nonlinear differential equations or systems with singular restoring forces occur naturally in the description of particules submitted to Newtonian type forces or to restoring forces caused by compressed gases. For example, the differential equation

$$x'' + cx' + \frac{x}{1-x} = e(t)$$

described the motion of a piston in a cylinder closed at one extremity and submitted to a time-periodic exterior force $e(t)$ with mean value

$$\bar{e} =: \frac{1}{T} \int_0^T e(t)\, dt$$

equal to zero, the restoring force of a perfect gas and a viscosity friction. It was considered, in the early sixties, by FORBAT and his students [40], [60], TAMPIERI (see [28]), DERWIDUÉ ([25],[26],[27],[28]) (who proposed to call it *Forbat's equation*), MÜLLER ([104],[105]), and finally by FAURE ([34], [35], [36]) who used, in contrast to the previous works based upon dissipativeness properties, the Leray-Schauder method to study the equation

$$x'' + cx' + \frac{x}{x+k} = e(t),$$

for negative values of k.

Critical point theory was then first applied to Hamiltonian systems with singularities by GORDON ([51], [52]) in the mid seventies, and many contributions in this direction have been given recently in the unforced case. The use of topological methods in forced equations was renewed by LAZER and SOLIMINI [74] who considered problems suggested by the two fondamental examples

$$x'' + \frac{1}{x^\alpha} = h(t), \tag{70}$$

and

$$x'' - \frac{1}{x^\alpha} = h(t), \tag{71}$$

for which a necessary condition for the existence of a positive T-periodic solution is respectively that

$$\int_0^T h(t)\, dt > 0 \text{ and } \int_0^T h(t)\, dt < 0,$$

respectively, as shown by integrating both members of the equation between 0 and T. LAZER and SOLIMINI have shown that those conditions are also sufficient (if, in the second case, one assumes moreover that $\alpha \geq 1$) by using respectively upper and lower solutions and a combination of truncation techniques and Schauder fixed point theorem. Generalizations to the case of systems of the form

$$x'' + \nabla F(x) = h(t)$$

haven been given by SOLIMINI [128] using critical point theory although some extensions of the Lazer-Solimini result, including situations where dissipation is present, have been given by HABETS and SANCHEZ ([54], [55]) using upper and lower solutions and coincidence degree arguments. We shall describe here slight variants of the Lazer-Solimini and some of the Habets-Sanchez results and obtain moreover associated multiplicity results using bifurcation from infinity techniques inspired from [97], [98], [94].

6.2 The case of an attractive type restoring force

In this section we shall consider the differential equation

$$x'' + f(x)x' + g(x) = h(t), \tag{72}$$

where $f :]0, +\infty[\to \mathbf{R}$, $g :]0, +\infty[\to \mathbf{R}$ and $h : [0, T] \to \mathbf{R}$ are continuous. The *method of upper and lower solutions* provides the following existence theorem for the T-periodic solutions of equation (72), which can be proved by topological degree arguments (see e.g. [48], [90], [91]).

Lemma 6.1. *Assume that there exist T-periodic C^2-functions α and β such that $\alpha(t) \leq \beta(t)$ and*

$$\alpha''(t) + f(\alpha(t))\alpha'(t) + g(\alpha(t)) \geq h(t),$$

$$\beta''(t) + f(\beta(t))\beta'(t) + g(\beta(t)) \leq h(t),$$

for all $t \in [0, T]$. Then equation (72) has at least one T-periodic solution x with $\alpha(t) \leq x(t) \leq \beta(t)$ for all $t \in [0, T]$.

The function α (resp. β) is called a *lower solution* (resp. an *upper solution*) of the T-periodic problem for (72).

We shall also need the following result which is proved in [88] using degree arguments (see also [96]).

Lemma 6.2. *For each $C \in \mathbf{R}$ and each continuous function $e : [0, T] \to \mathbf{R}$ such that $\bar{e} = 0$, the equation*

$$y'' + f(C + y)y' = e(t)$$

has at least one T-periodic solution y such that

$$\bar{y} = 0 \text{ and } \|y\|_\infty \leq K\|e\|_{L^2},$$

for some K independent of C and u.

We can now prove the following result, which is due to LAZER-SOLIMINI ([74]) for $f = 0$ and is a special case of a result of HABETS-SANCHEZ ([55]) for arbitrary f.

Theorem 6.1. *Assume that the function $g :]0, +\infty[\to \mathbf{R}$ is such that the following conditions hold.*
1. $g(x) \to +\infty$ as $x \to 0+$.
2. $\limsup_{x \to +\infty} [g(x) - \bar{h}] < 0$.
Then equation (72) has at least one T-periodic positive solution.

Proof. We first observe that, by assumption 1, there exist a constant $\alpha > 0$ such that

$$g(\alpha) \geq h(t)$$

for all $t \in [0, T]$, and hence α is a lower solution for (72) with T-periodic boundary conditions. We now write $h(t) = \bar{h} + \tilde{h}(t)$. By assumption 2, there exists $R > 0$ be such that $g(x) \leq \bar{h}$ for $x \geq R$ and, using Lemma 6.2, let y be a T-periodic solution of the equation

$$y'' + f(C + y)y' = \tilde{h}(t)$$

with C sufficiently large so that $C + y(t) \geq \max(\alpha, R)$ for all $t \in [0, T]$. Letting $\beta(t) = C + y(t)$, we have

$$\beta''(t) + f(\beta(t))\beta'(t) + g(\beta(t)) =$$

$$y''(t) + f(C + y(t))y'(t) + g(C + y(t)) \leq \tilde{h}(t) + \bar{h} = h(t),$$

so that $\beta(t) \geq \alpha$ is an upper solution for (72) with T-periodic boundary conditions. The existence of a T-periodic solution x of (72) with $\alpha \leq x(t) \leq \beta(t)$ follows then from Lemma 6.1. ■

Corollary 6.1. *Assume that the function* $g :]0, +\infty[\rightarrow]0, +\infty[$ *is such that the following conditions hold.*
1. $g(x) \rightarrow +\infty$ *as* $x \rightarrow 0+$.
2. $\lim_{x \rightarrow +\infty} g(x) = 0$.
Then equation (72) has a positive T-periodic solution if and only if $\bar{h} > 0$.

Proof. The necessity follows immediately by integrating the equation over $[0, T]$ and using the positivity of g. The sufficiency follows from Theorem 6.1. ∎

It is clear that Corollary 6.1 contains as special case the Lazer-Solimini result for equation (70). It can also be applied to a generalized differential equation of Forbat type

$$u'' + j(u)u' + k\frac{u}{1-u} = e(t), \tag{73}$$

when $j : \mathbf{R} \rightarrow \mathbf{R}$ is continuous and $e : [0, T] \rightarrow \mathbf{R}$ is continuous and has mean value zero. The search of T-periodic solutions u of (73) such that $u(t) < 1$ for all $t \in [0, T]$ corresponds, by the change of variable $1 - u = x$ to that of the positive T-periodic solutions of the equation

$$x'' + j(1-x)x' - \frac{k}{x} = -k + e(t),$$

which is of type (72) with

$$f(x) = j(1-x),\ g(x) = -\frac{k}{x},\ h(t) = -k + e(t).$$

Consequently, Theorem 6.1 implies that equation (73) has at least one T-periodic solution u such that $u(t) < 1$ for all $t \in [0, T]$ when $k < 0$. This is the situation considered in [34], [35] and [36].

6.3 The case of repulsive type restoring force

We now consider the equation

$$x'' + cx' - g(x) = h(t), \tag{74}$$

where $c \in \mathbf{R}$, $g :]0, +\infty[\rightarrow \mathbf{R}$ is continuous and $h : [0, T] \rightarrow \mathbf{R}$ is Lebesgue integrable. To prove the existence of a positive T-periodic solution to (74), we shall use the following continuation theorem which is a special case of Theorem 2.4.

Lemma 6.3. *Assume that there exist* $R > r > 0$ *such that the following conditions hold.*
1. For each $\lambda \in]0, 1]$, *each possible T-periodic positive solution of the equation*

$$x'' + cx' - \lambda g(x) = \lambda h(t), \tag{75}$$

satisfies the inequality $r < x(t) < R$ *for all* $t \in [0, T]$.
2. Each possible positive solution of the equation in $\mathbf{R}$

$$g(a) + \bar{h} = 0 \tag{76}$$

satisfies the inequality $r < a < R$.
3. $[g(r)+\bar{h}][g(R)+\bar{h}] < 0$.
Then equation (74) has at least one solution x *such that* $r < x(t) < R$ *for all* $t \in [0, T]$.

We now prove the following existence result for the T-periodic solutions of equation (74).

Theorem 6.2. *Assume that the function* $g :]0, +\infty[\to \mathbb{R}$ *is such that the following conditions hold.*
1. $g(x) \geq -ax - b$ *for some* $a \in]0, (2T^2 \exp 2|c|T)^{-1}[$, $b \geq 0$ *and all* $x > 0$.
2. $\lim_{x \to 0+} g(x) = +\infty$.
3. $\limsup_{x \to +\infty} [g(x) + \bar{h}] < 0$.
4. $\int_0^1 g(x)\, dx = +\infty$.
Then equation (74) has at least one T-periodic solution.

Proof. We show that the conditions of Lemma 3 are satisfied.
a) *a trick of* WARD ([139])
We first notice that assumption 1 implies that

$$|g(u)| \leq g(u) + 2au + 2b \tag{77}$$

for all $u > 0$. If x is a possible solution of (75) for some fixed $\lambda \in]0, 1]$, we have

$$\frac{1}{T}\int_0^T g(x(t))\, dt + \bar{h} = 0, \tag{78}$$

and

$$\int_0^T |x''(t) + cx'(t)|\, dt \leq \lambda \int_0^T |g(x(t))|\, dt + \lambda \int_0^T |h(t)|\, dt,$$

so that, using (77), we get

$$\|x'' + cx'\|_{L^1} \leq \lambda[\int_0^T g(x(t))\, dt + 2a\|x\|_{L^1} + 2bT + \|h\|_{L^1} - \tag{79}$$

$$2\lambda[a\|x\|_{L^1} + bT + \|h^-\|_{L^1}].$$

b) *a trick of* VILLARI ([135])
We first notice that assumption 2 implies the existence of $R_0 > 0$ such that

$$g(u) > 0 \text{ and } g(u) + \bar{h} > 0$$

whenever $0 < u \leq R_0$ and assumption 3 implies the existence of $R_1 > R_0$ such that

$$g(u) + \bar{h} < 0$$

whenever $u \geq R_1$. Consequently, if $0 < x(t) \leq R_0$ for all $t \in [0, T]$, we get $g(x(t)) + \bar{h} > 0$ for those t and hence

$$\frac{1}{T}\int_0^T g(x(t))\, dt + \bar{h} > 0,$$

a contradiction to (9). Thus, there exists t_0 such that $x(t_0) > R_0$. Similarly, if $x(t) \geq R_1$ for all $t \in [0, T]$, then $g(x(t)) + \bar{h} < 0$ for those t and

$$\frac{1}{T}\int_0^T g(x(t))\,dt + \bar{h} < 0,$$

a contradiction to (78). Thus there exist t_1 such that $x(t_1) < R_1$.

c) *a combination of a and b*

Notice first that as $\int_0^T x'(t)\,dt = 0$, there exists t_3 such that $x'(t_3) = 0$. From this observation and the identity

$$\exp(ct)[x''(t) + cx'(t)] = [\exp(ct)x'(t)]',$$

we get

$$\exp(ct)x'(t) = \int_{t_3}^{t} \exp(cs)[x''(s) + cx'(s)]\,ds,$$

and hence, using the results of b,

$$x(t) = x(t_1) + \int_{t_1}^{t} \exp(-c\tau) \int_{t_3}^{\tau} \exp(cs)[x''(s) + cx'(s)]\,ds\,d\tau$$
$$< R_1 + T\exp(2|c|T)\|x'' + cx'\|_{L^1}$$

so that

$$\|x\|_{L^1} < R_1 T + T^2 \exp(2|c|T)\|x'' + cx'\|_{L^1}. \tag{80}$$

Combining (80) with (79), we get

$$[1 - 2aT^2\exp(2|c|T)]\|x'' + cx'\|_{L^1} < 2\lambda[bT + \|h^-\|_{L^1}],$$

and hence,

$$\|x'' + cx'\|_{L^1} < \lambda R_2,$$

for some $R_2 > 0$ independent of x and λ, which gives then

$$|x'(t)| < \lambda \exp(2T|c|)R_2 = \lambda R_3, \tag{81}$$

and

$$x(t) < R_1 + T\exp(2|c|T)R_2 = R, \tag{82}$$

for all $t \in [0, T]$.

d) *a trick of* LAZER-SOLIMINI ([74])

We get now, from equation (75),

$$x''x' + cx'^2 - \lambda g(x)x' = \lambda h x',$$

and hence

$$\frac{(x'(t))^2}{2} - \frac{(x'(t_0))^2}{2} + c\int_{t_0}^{t} (x'(s))^2\,ds - \lambda \int_{t_0}^{t} g(x(s))x'(s)\,ds =$$

$$\lambda \int_{t_0}^{t} h(s)x'(s)\,ds,$$

where t_0 comes from (b). Consequently, we have

$$\lambda \int_{x(t)}^{x(t_0)} g(s)ds \le \frac{(x'(t_0))^2}{2} + |c| \int_0^T (x'(s))^2\,ds + \lambda \int_{t_0}^{t} |h(s)x'(s)|\,ds$$

$$\le \lambda^2[(\frac{1}{2} + |c|T)R_3^2 + \|h\|_{L^1} R_3],$$

so that

$$\int_{x(t)}^{x(t_0)} g(s)\,ds \le R_4, \tag{83}$$

for some $R_4 > 0$ independent of x and λ. Therefore, if t is such that $x(t) \le R_0$, we deduce from (83) that

$$\int_{x(t)}^{R_0} g(s)\,ds + \int_{R_0}^{x}(t_0) g(s)\,ds \le R_4,$$

and hence

$$\int_{x(t)}^{R_0} g(s)\,ds \le R_4 + \int_{R_0}^{R} |g(s)|\,ds = R_5. \tag{84}$$

Now, by assumption 4, there exist $r \in]0, R_0[$ such that

$$\int_{r}^{R_0} g(s)\,ds > R_5,$$

so that the monotonicity of the function $\int_{\cdot}^{R_0} g(s)\,ds$ and the inequality (84) imply that $x(t) > r$. Thus either $x(t) > R_0$ or $x(t) > r$, so that $x(t) > r$ for all $t \in [0, T]$, and assumption 1 of Lemma 6.3 is satisfied. If we notice that, by construction $r < R_0$ and $R > R_1$, it follows from section b of the proof that the assumptions 2 and 3 of Lemma 6.3 also hold, and the proof is complete. ■

Corollary 6.2. *Assume that the function $g :]0, +\infty[\to]0, +\infty[$ is such that the following conditions hold.*
1. $g(x) \to +\infty$ as $x \to 0+$.
2. $\lim_{x \to +\infty} g(x) = 0$.
3. $\int_0^1 g(s)\,ds = +\infty$.
Then equation (74) has a positive T-periodic solution if and only if $\bar{h} < 0$.

Proof. The necessity follows immediately from the positivity of g by integrating the equation over $[0, T]$. The sufficiency follows from Theorem 6.2. ■

It is clear that Corollary 6.2 contains as special case the Lazer-Solimini result for equation (71). It can also be applied to a differential equation of Forbat type

$$u'' + cu' + k\frac{u}{1-u} = e(t), \tag{85}$$

when $e : [0,T] \to \mathbf{R}$ is Lebesgue integrable and has mean value zero. The search of T-periodic solutions u of (85) such that $u(t) < 1$ for all $t \in [0,T]$ corresponds, by the change of variable $1 - u = x$ to that of the positive T-periodic solutions of the equation

$$x'' + cx' - \frac{k}{x} = -k + e(t), \tag{86}$$

which is of type (74) with

$$f(x) = c, \; g(x) = -\frac{k}{x}, \; h(t) = -k + e(t).$$

Consequently, Theorem 6.2 implies that equation (85) has at least one T-periodic solution u such that $u(t) < 1$ for all $t \in [0,T]$ when $k > 0$. This is the situation considered in [40], [60], [25], [26], [27], [28], [104], [105].

6.4 Bifurcation from infinity and multiplicity of the solutions

Let $X = C([0,T],\mathbf{R})$, $D(L) = \{x \in C([0,T]) : x \in C^1, x(0) = x(T),\; x'(0) = x'(T)\}$, $L : D(L) \subset X \to X, x \mapsto -x'' - cx'$, $\Omega_{r,R} = \{x \in X : r < x(t) < R,\; t \in [0,T]\}$, $N : \overline{\Omega_{r,R}} \to X,\; x \mapsto G(x(.)) + h(.)$, where the continuous functions $G :]0,+\infty[\to [0,+\infty[$ and $h \in L^1([0,T])$ satisfy the following conditions
(H_1) $\lim_{s \to 0+} G(s) = +\infty$.
(H_2) $\lim_{s \to +\infty} G(s) = 0$.
(H_3) $\bar{h} < 0$.

We shall be interested in the structure of the set $(\mu, x) \in \mathbf{R} \times X$ of the solutions with x positive of the nonlinear equation

$$Lx - \mu x + Nx = 0, \tag{87}$$

near $\mu = 0$. The following result, adapted from [89] and which uses ideas from [97], [98], provides multiplicity results for the solutions of equation (87), i.e. for the positive T-periodic solutions of the equation

$$x'' + cx' + \mu x = g(x) + h(t). \tag{88}$$

Theorem 6.3. *Assume that conditions (H_1), (H_2) and (H_3) are satisfied. Then there exists $\eta > 0$ such that the following holds :*
(a) equation (88) has at least one positive solution x for $0 \leq \mu \leq \delta$);
(b) equation (88) has at least two positive T-periodic solutions x for $-\eta \leq \mu < 0$).

Proof. We first notice that equation (88) is of the form (74) with $g(x) = G(x) - \mu x$, satisfying the following conditions (with $\delta > 0$ fixed)
(H_1') $\lim_{x \to 0+} g(x) = +\infty$.
(H_2') $g(x) \geq -\mu_0 x$ for all $x > 0$ and $\mu \leq \delta$.
(H_3') $\limsup_{x \to +\infty} [g(x) + \bar{h}] < 0$ for $\mu \geq 0$.

Therefore the results of Theorem 6.2 are valid for equation (88) when $\mu \in [0,\delta]$ and the arguments of the proof show that, for those values of μ, all possible positive solutions are contained in $\Omega_{r,R}$ with r and R given by Theorem 6.2 and the coincidence degree $D_L(L-\mu I-N,\Omega_{r,R})$ of $L-\mu I-N$ on the open bounded set $\Omega_{r,R}$ is well defined and equal to one in absolute value; hence there exists $\gamma>0$ such that the same is true for $-\gamma\le\mu\le\delta$. Consequently, there exists a continuum $\mathcal{C}_R$ of solutions (μ,x) of (17) in $[-\gamma,\delta]\times\Omega_{r,R}$ whose projection on $\mathbf{R}$ is $[-\gamma,\delta]$. On the other hand, the fundamental theorem on bifurcation from infinity (see e.g. [120]) implies the existence of a continuum $\mathcal{C}_\infty$ of positive solutions (μ,x) bifurcating from infinity at $\mu=0$. More explicitely, there exists $\alpha>0$ such that for each $0<\epsilon\le\alpha$ there is a subcontinuum $\mathcal{C}_\epsilon\subset\mathcal{C}_\infty$ of positive solutions contained in $U_\epsilon(0,\infty)=\{(\mu,x)\in\mathcal{C}_\infty:|\mu|<\epsilon,\|x\|>1/\epsilon\}$, and connecting $(0,\infty)$ to $\partial U_\epsilon(0,\infty)$. Necessarily, for $\epsilon=\min(1/R,\gamma,\alpha)$, we have $\mathcal{C}_\epsilon\subset\{(\mu,x)\in\mathcal{C}_\infty:-\epsilon\le\mu<0\}$, and hence we obtain a second solution with $\|x\|>R$ for $-\eta=-\min(\epsilon,\beta)\le\mu<0$, with $\beta=\sup\{-\mu:(\mu,x)\in\mathcal{C}_\epsilon\}$. ∎

7 Stability and index of periodic solutions

7.1 Introduction

Topological degree techniques have been widely applied in the study of the existence and the multiplicity of solutions of nonlinear boundary value problems, and in particular of periodic solutions of nonlinear differential equations. Much less has been done using those techniques in the study of the *stability* of those solutions. Concentrating on the periodic case, we can quote some pioneering studies for arbitrary systems ([63]), [95]) and for second order equations ([105], [62], [83]). Among the more recent results, we may quote [92] (for periodic solutions of first order scalar equations), [111], [112], [113], [114], [115], [72], [73]. We shall analyze briefly here some of the results of ORTEGA taken from [111] and [112]. The reader can consult the original papers for more details and further developments.

7.2 Planar periodic systems

We consider the following two-dimensional system

$$x'=f(t,x),\tag{89}$$

where the continuous mapping $f:\mathbf{R}\times\mathbf{R}^2\to\mathbf{R}$ is T-periodic with respect to t for fixed x and of class C^1 with respect to x for fixed t. We denote by $x(t;x_0)$ the solution of (89) such that $x(0;x_0)=x_0$ and introduce the *Poincaré map* $P_T:x_0\mapsto x(T;x_0)$. This map is defined for the (open) subset of $\mathbf{R}^2$ made of the x_0 such that $x(.;x_0)$ is defined over $[0,T]$. It is well know then that $x(.;x_0)$ will be a T-periodic solution of (89) if and only if x_0 is a fixed point of P_T. If x is an *isolated* T-periodic solution of (89), then $x(0)$ is an isolated fixed point of P_T and hence the Brouwer index

$$\mathrm{ind}[P_T,x(0)]:=\deg(I-P_T,B_r(x(0)),0),$$

where $r>0$ is sufficiently small, is well defined and called the *index of period* T of x and denoted by $\gamma_T(x)$. Similarly, we shall have to consider the mapping $P_{2T}:x_0\mapsto x(2T;x_0)$

and if x is an isolated 2T-periodic solution of (89) we shall define the *index of period* $2T$ of x by

$$\gamma_{2T}(x) = \mathrm{ind}[P_{2T}, x(0)] = \deg(I - P_{2T}, B_r(x(0)), 0).$$

Notice that every T-periodic solution x of (89) is also a 2T-periodic solution and if it is isolated as a 2T-periodic solution, it is also isolated as a T-periodic solutions and admits both indices γ_T and γ_{2T}.

Definition 7.1. *A T-periodic solution x of (89) will be called* nondegenerate of period T (resp. of period 2T) *if the linearized equation*

$$y' = f_x(t, x(t))y \tag{90}$$

has no nontrivial T-periodic (resp. 2T-periodic) solution.

If x is T-periodic and nondegenerate of period $2T$, then an implicit function argument shows that x is isolated for periods T and $2T$ and the linearization formula for degree together with the Leray-Schauder formula for linear mappings show that

$$|\gamma_T(x)| = |\gamma_{2T}(x)| = 1.$$

Let us assume from now on that

$$\mathrm{div}\ f(t, x) < 0, \tag{91}$$

for all (t, x).

Theorem 7.1. *Let x be a T-periodic solution of (89) which is nondegenerate of period $2T$. Then $\gamma_{2T}(x) = 1 (resp. -1)$ if and only if x is uniformly asymptotically stable (resp. is unstable).*

Proof. Let μ_1 and μ_2 be the characteristic multipliers of the variational equation (90) considered as an equation with T-periodic coefficients. Recall that if $Y(t)$ is the fundamental matrix of (90), then μ_1 and μ_2 are the eigenvalues of the matrix $Y(T)$, which implies that the identity

$$\mu^2 - (\mu_1 + \mu_2)\mu + \mu_1\mu_2 = \det[Y(T) - \mu I],$$

for all $\mu \in \mathbf{C}$, and the fact that $\mu_2 = \overline{\mu_1}$ if μ_1 is complex. In particular, using Liouville's formula and assumption (91), we have

$$\mu_1\mu_2 = \det Y(T) = \exp\left(\int_0^T \mathrm{div}\ f_x(s, x(s))\, ds \in]0, 1[. \tag{92}$$

The nondegeneracy condition implies that $\mu_i \notin \{-1, 1\}$, $(i = 1, 2)$. Condition (92) implies that if say μ_1 is not real, then $\mu_2 = \overline{\mu_1}$ and $|\mu_1|^2 = |\mu_2|^2 = \mu_1\mu_2 < 1$. Thus, in any case,

$$|\mu_i| \neq 1, \ (i = 1, 2). \tag{93}$$

Now, for each $x_0 \in \mathbf{R}^2$, we have

$$P_{2T}(x_0) = x(2T; x_0) = x[T; x(T; x_0)] = P_T(P_T(x_0)),$$

and hence

$$P'_{2T}(x_0) = P'_T(P_T(x_0)) \circ P'_T(x_0),$$

which implies in particular, as x is T-periodic, that

$$P'_{2T}[x(0)] = P'_T[x(0)] \circ P'_T[x(0)] \tag{94}$$

By definition of P_T we have

$$P'_T(x_0) = \frac{\partial x}{\partial x_0}(T; x_0),$$

and from the identities

$$x'(t; x_0) = f[t, x(t; x_0)], x(0; x_0) = x_0,$$

we get

$$\frac{d}{dt}[\frac{\partial x}{\partial x_0}(t; x_0)] = f'_x[t, x(t; x_0)]\frac{\partial x}{\partial x_0}(t; x_0), \; \frac{\partial x}{\partial x_0}(0; x_0) = I.$$

Consequently,

$$Y(t) = \frac{\partial x}{\partial x_0}[0; x(0)],$$

and hence

$$P'_T[x(0)] = \frac{\partial x}{\partial x_0}[T; x(0)] = Y(T).$$

As a consequence,

$$\det[I - P'_{2T}(x(0))] = \det\{[I - P'_T[x(0)] \circ P'_T[x(0)]\} =$$

$$\det[I - Y(T) \circ Y(T)] = (1 - \mu_1^2)(1 - \mu_2^2).$$

and

$$\gamma_{2T}(x) = \operatorname{sign}[(1 - \mu_1^2)(1 - \mu_2^2)]. \tag{95}$$

Thus, if one of the μ_i (and hence the other one too) is not real, we have $|\mu_i| < 1, \; i = 1, 2,$ and

$$\gamma_{2T}(x) = \operatorname{sign}|(1 - \mu_1^2)|^2 = 1.$$

Now, x is uniformly asymptotically stable if and only if $|\mu_i| < 1, \; (i = 1, 2)$, and hence, in the case of non real multipliers, the statement about the uniform stability is proved. If the multipliers are real, then $\gamma_{2T} = 1$ if and only if μ_1^2 and μ_2^2 are both smaller or both larger than one, and the second case is excluded by relation (92), so that the statement about the uniform stability is proved also in this case. Finally, x is unstable if and only if one multiplier, say μ_1 has absolute value greater than one, which can only arrive if they are real. Thus, this is equivalent to the fact that $\mu_1^2 > 1$ so that necessarily, by relation (92), $\mu_2^2 < 1$ and, by (95), those two conditions are equivalent to $\gamma_{2T} = -1$. ∎

7.3 The case of a second order differential equation

We shall consider in this section the second order differential equation

$$u'' + cu' + g(t, u) = 0, \tag{96}$$

with $c > 0$ and $g : \mathbf{R} \times \mathbf{R} \to \mathbf{R}$ a continuous function, T-periodic with respect to the first variable and having a continuous partial derivative with respect to the second variable. Such an equation is equivalent to the planar system

$$x_1' = x_2,\ x_2' = -cx_2 - g(t, x_1), \tag{97}$$

This system satisfies condition (91) as

$$\text{div}\,(x_2, -cx_2 - g(t, x_1)) = -c < 0$$

for all $(x_1, x_2) \in \mathbf{R}^2$. The aim of this section is to characterize the stability of a T-periodic solution u of (96) (i.e. of the associated T-periodic solution $x = (u, u')$ of (97)) in terms of its index $\gamma_T(u) := \gamma_T(x)$ only. If α and β are real functions defined over $[0, T]$, we shall write $\alpha \ll \beta$ if $\alpha(t) \leq \beta(t)$ for all $t \in [0, T]$ and the strict inequality holds on a subset of positive Lebesgue measure.

Lemma 7.1. *Assume that $c > 0$ and that the real function α is continuous and T-periodic and such that*

$$\alpha \ll \frac{\pi^2}{T^2} + \frac{c^2}{4}.$$

Then the linear equation

$$v'' + cv' + \alpha(t)v = 0 \tag{98}$$

does not admit negative characteristic multipliers.

Proof. Setting

$$v(t) = \exp(-\frac{c}{2}t)w(t),$$

equation (98) becomes

$$w'' + [\alpha(t) - \frac{c^2}{4}]w = 0 \tag{99}$$

and the corresponding solutions v and w obviously have the same zeros. If (98) admits a negative characteristic multiplier μ, then it has a nontrivial solution v such that

$$v(t + T) = \mu v(t)$$

for all $t \in \mathbf{R}$. If t is such that $v(t) \neq 0$, then v has opposite signs at t and $t + T$ and therefore must vanish at some $t_0 \in]t, t+T[$, and hence also at $t_0 + T$. Thus w is a solution of equation (99) which satisfies the two-point boundary conditions

$$y(t_0) = y(t_0 + T) = 0.$$

But $\frac{\pi^2}{T^2}$ is the first eigenvalue of the two-point boundary value problem on an interval of length T for the operator $-w''$ and Lemma 3 of [99] and the assumption over α imply that w must be identically zero, a contradiction. ∎

Theorem 7.2. *Assume that u is an isolated T-periodic solution of (96) such that the condition*

$$\frac{\partial g}{\partial u}(t, u(t)) \leq \frac{\pi^2}{T^2} + \frac{c^2}{4} \tag{100}$$

holds for all $t \in \mathbf{R}$. Then u is uniformly asymptotically stable (resp. unstable) if and only if $\gamma_T(u) = 1$ (resp. $\gamma_T(u) = -1$).

Proof. If $\frac{\partial g}{\partial u}(t.u(t)) = \frac{\pi^2}{T^2} + \frac{c^2}{4}$ for all $t \in \mathbf{R}$, then the variational equation

$$v'' + cv' + \frac{\partial g}{\partial u}(t, u(t))v = 0, \tag{101}$$

has constant coefficients and the characteristic exponents have negative real parts, which immediately implies that u is uniformly asymptotically stable and $\gamma_T(u) = \gamma_{2T}(u) = 1$. We can therefore assume that

$$\frac{\partial g}{\partial u}(., u(.)) \ll \frac{\pi^2}{T^2} + \frac{c^2}{4}.$$

Lemma 7.1 implies then that the variational equation (101) has no negative characteristic multipliers. We already know from the proof of Proposition 7.1 that $\gamma_T(u) = \gamma_{2T}(u) = 1$ and u is uniformly asymptotically stable if the characteristic multiplies are not real. If the μ_i are real, they are both positive and

$$\text{sign}[(1 - \mu_1^2)(1 - \mu_2^2)] = \text{sign}[(1 - \mu_1)(1 - \mu_2)(1 + \mu_1)(1 + \mu_2)] =$$

$$\text{sign}[(1 - \mu_1)(1 - \mu_2)],$$

so that $\gamma_T(u) = \gamma_{2T}(u)$ and the result follows from Theorem 7.1. ∎

Remark 7.1. The following example from [111] shows that the assumption in Theorem 7.2 is optimal. Let β be a T-periodic function defined by

$$\beta(t) = -\omega_1^2 + \frac{c^2}{4}, \ t \in]0, a[,$$

$$\beta(t) = \omega_2^2 + \frac{c^2}{4}, \ t \in]a, T[,$$

for some $0 < a < T$ and ω_1, $\omega_2 > 0$. We consider the equation

$$u'' + cu' + \omega(t)u = 0.$$

If the assumption of Theorem 7.2 does not hold for this equation, then $\omega_2^2 > \frac{\pi^2}{T^2}$ and we can fix a so that $\omega_2(T - a) = \pi$. Direct integration of the equation shows that one can select ω_1 in such a way that the trivial solution is unstable and has index 1. Hence, $\frac{\pi^2}{T^2} + \frac{c^2}{4}$ cannot be replaced by any greater constant. One could object that in this example the function ω is not continuous but this can be of course easily overcame by an approximation argument.

Remark 7.2. Using a class of systems introduced by SMITH [127] and for which the Massera's convergence theorem holds, ORTEGA [113] has obtained the following partial improvement of Theorem 7.2 :

Proposition 7.1. *Let u be a T-periodic solution of (96) such that*

$$\frac{\partial g}{\partial u}(t, u(t)) < \frac{c^2}{4}$$

for all $t \in \mathbf{R}$. Then the conclusion of Theorem 7.2 holds.

Remark 7.3. Using a combination of degree theory with the theory of center manifolds, ORTEGA [114] has obtained the following improvement of Theorem 7.1 :

Proposition 7.2. *Let x be a T-periodic solution of (89) which is isolated of period $2T$. Then $\gamma_{2T}(x) = 1$(resp. $\neq 1$) if and only if x is uniformly asymptotically stable (resp. is unstable).*

He has also obtained a version of this result in $\mathbf{R}^n$.

7.4 Second order equations with convex nonlinearity

We consider in this section the second order parametric equation

$$u'' + cu' + g(t, u) = s, \tag{102}$$

where $c > 0$, s is a real parameter and the continuous function $g : \mathbf{R} \times \mathbf{R} \to \mathbf{R}$ is T-periodic with respect to the first variable and having a continuous partial derivative with respect to the second variable. Assume moreover that g is strictly convex, namely that

$$[\frac{\partial g}{\partial u}(t, u) - \frac{\partial g}{\partial u}(t, v)](u - v) > 0, \tag{103}$$

for all $t \in \mathbf{R}$ and all $u \neq v$ in $\mathbf{R}$. To insure that condition (100) holds, we shall assume also that

$$\lim_{u \to +\infty} \frac{\partial g}{\partial u}(t, u) \leq \frac{\pi^2}{T^2} + \frac{c^2}{4}. \tag{104}$$

Finally, let us introduce the coercitivity condition

$$\lim_{|u| \to \infty} g(t, u) = +\infty \tag{105}$$

introduced in [33]. Following [111], we shall complete the existence discussion given in [33] by results on the stability of the corresponding T-periodic solutions. The following lemma will be used in the proof.

Lemma 7.2. *Consider the differential operator L_α defined by*

$$L_\alpha(u) = u'' + cu' + \alpha(.)u,$$

acting on T-periodic functions, where $\alpha \in L^1(\mathbf{R}/T\mathbf{Z})$ satisfies condition

$$\alpha \ll \frac{\pi^2}{T^2} + \frac{c^2}{4}. \tag{106}$$

Then the following conclusions are true.
(a) For each real μ each possible T-periodic solution u of the equation $L_\alpha(u) = \mu$ is either trivial or different from zero for each $t \in [0, T]$.
(b) Let α_1 and α_2 be functions in $L^1(\mathbf{R}/T\mathbf{Z})$ satisfying (106) and such that $\alpha_1 \ll \alpha_2$. Then the equations

$$L_{\alpha_i}(u) = 0, \ (i = 1, 2),$$

cannot admit nontrivial T-periodic solutions simultaneously.

Proof. If a nontrivial T-periodic solution u of equation $L_\alpha(u) = \mu$ vanishes at some t_0, then it vanishes also at $t_0 + T$, and, by the same reasoning as in the proof of Lemma 7.1, the function w defined by $w(t) = \exp(-\frac{c}{2}t)u(t)$ is a solution of the differential equation

$$w'' + [\alpha(t) - \frac{c^2}{4}]w = 0$$

which vanishes at t_0 and $t_0 + T$, a contradiction with Lemma 3 of [99]. To prove (b), we first notice that conclusion (a) obviously holds also for the adjoint operator L^*_α defined by $L^*_\alpha(u) = u'' - cu' + \alpha(.)u$. If both equations $L_{\alpha_i}(u) = 0$ admit nontrivial T-periodic solutions u_1 and u_2 respectively, then the adjoint equation $L^*_{\alpha_1}(u) = 0$ admits a nontrivial T-periodic solution φ. By the part (a) of the Lemma, we can assume without loss of generality that $u_2 > 0$ and $\varphi > 0$ on $[0, T]$. Moreover,

$$L_{\alpha_1}(u_2) = L_{\alpha_2}(u_2) + (\alpha_1 - \alpha_2)u_2 = (\alpha_1 - \alpha_2)u_2,$$

so that, using the Fredholm alternative, we get

$$0 = \int_0^T [\alpha_1(t) - \alpha_2(t)]u_2(t)\varphi(t)\, dt,$$

which is a contradiction with the assumption $\alpha_1 \ll \alpha_2$. ■

We shall use also the following truncation of equation (102). Given $u_+ > 0 > u_-$, define the function (bounded and Lipschitzian in u) g^* by

$$g^*(t, u) = g(t, u), \ (u \in]u_-, u_+[);$$
$$g^*(t, u) = g(t, u_-), \ (u \leq u_-), \ g^*(t, u) = g(t, u_+), \ u \geq u_+.$$

and the associated truncated equation

$$u'' + cu' + g^*(t, u) = s. \tag{107}$$

The following lemma is essentially proved like Lemma 2 of [33].

Lemma 7.3. *Assume that condition (105) holds. Then the following conclusions are valid.*
(a) There exists an increasing function $M(s)$ such that each possible T-periodic solution of (102) satisfies the inequality

$$|u(t)| + |u'(t)| < M(|s|), \ (t \in \mathbf{R}).$$

(b) Given $S > 0$ and $u_+ > M(S)$, $u_- < -M(S)$, the set of possible T-periodic solutions of (102) and (107) coincide when $|s| \leq S$ for $|u_\pm|$ sufficiently large.

We shall also need the following preliminary result.

Lemma 7.4. *Assume that the function g satisfies conditions (103) and (104). Then, given s_1, $s_2 \in \mathbf{R}$ and u_1, u_2 be T-periodic solutions of (102) for $s = s_1$, $s = s_2$ respectively such that $u_1 \neq u_2$. Then either $u_1 > u_2$ or $u_1 < u_2$ on $[0, T]$.*

Proof. The difference $v = u_1 - u_2$ satisfies an equation of the form

$$L_\alpha(v) = s_1 - s_2,$$

for some T-periodic function α which, because of conditions (103) and (104) satisfies the conditions of Lemma 7.2. The conclusion follows from this lemma. ■

We can now state and prove the main result of this section.

Theorem 7.3. *Assume that the function g satisfies conditions (100), (103) and (104). Then there exists $s_0 \in \mathbf{R}$ such that the following conclusions hold.*
(i) If $s > s_0$, equation (102) has exactly two T-periodic solutions, one uniformly asymptotically stable and another unstable.
(ii) If $s = s_0$, (102) has a unique T-periodic solution which is not asymptotically stable.
(iii) If $s < s_0$, every solution of (102) is unbounded.

Proof. It follows from the Ambrosetti-Prodi's type results of [33] that there exists $s_0 \in \mathbf{R}$ such that, for $s > s_0$ equation (102) has at least two T-periodic solutions, for $s = s_0$, (102) has at least one T-periodic solution and for $s < s_0$, (102) has no T-periodic solution. We now show that (102) has at most two T-periodic solutions, so that they are necessarily isolated. Otherwise, let u_i, $(i = 1, 2, 3)$ be three different T-periodic solutions. By Lemma 7.4, they can be ordered, say $u_1 < u_2 < u_3$. Setting $v_1 = u_2 - u_1$, $v_2 = u_3 - u_2$, we see immediately that v_i satisfies the equation $L_{\alpha_i}(v_i) = 0$ with $\alpha_i = [g(t, u_{i+1}) - g(t, u_i)]/v_i$, $(i = 1, 2)$. The strict convexity of g implies that $\alpha_1 < \alpha_2$ on $[0, T]$. Using Lemma 7.2, we obtain that either v_1 or v_2 must be zero, a contradiction.

Let now u_0 be a T-periodic solution of (102) for $s = s_0$. It follows from the continuity of the index that u_0 is degenerate and that $\gamma_T(u_0) = 0$, because, if it was not the case, equation (102) would have a T-periodic solutions for all $s \in]s_0 - \epsilon, s_0[$ for some $\epsilon > 0$. Notice now that if u_1 is another T-periodic solution of (102) for some $s_1 \geq s_0$, then, by Lemma 7.4, either

$$u_1 < u_0 \text{ and hence } \frac{\partial g}{\partial u}(t, u_1) < \frac{\partial g}{\partial u}(t, u_0) \text{ on } [0, T],$$

or

$$u_1 > u_0 \text{ and hence } \frac{\partial g}{\partial u}(t, u_1) > \frac{\partial g}{\partial u}(t, u_0) \text{ on } [0, T].$$

By the second part of Lemma 7.2, we conclude that u_1 is nondegenerate, and hence that $s_1 > s_0$. So we have proved that, for $s = s_0$ equation (102) has a unique T-periodic solution u_0 which satisfies $\gamma_T(u_0) = 0$, and hence cannot be uniformly asymptotically stable by Theorem 7.2.

A consequence of the above reasoning is the existence, for $s > s_0$, of exactly two T-periodic solutions of (102) and they are nondegenerate. Given $s_1 > s_0$, let us select $u_\pm$ so that the conclusion of Lemma 7.3 holds for $S = \max(|s_1|, |s_0|)$. By Lemma 7.2, the Poincaré operator $P^*_{T,s}$ associated to equation (107) (written as a planar system) has no fixed point on the boundary of the subset $\Omega = \{(x_1, x_2) : |x_1| + |x_2| < M_0\}$, with $M_0 = M(\max(|s_0|, |s_1|))$, for each $s \in [s_0, s_1]$. Hence, by the properties of Brouwer degree,

$$\deg[I - P^*_{T,s_1}, \Omega, 0] = \deg[I - P^*_{T,s_0}, \Omega, 0] = \gamma_T(u_0) = 0,$$

so that necessarily

$$\gamma_T(u_1) = -\gamma_T(u_1^*) = 1,$$

if u_1 and u_1^* denote the two nondegenerate T-periodic solutions of (102) (or of (107)) for $s = s_1$ (the index of such a T-periodic solution with respect to equation (102) or (107) is the same, being a local characteristic). Thus, by Theorem 7.2, (102) has exactly one uniformly asymptotically stable and one unstable T-periodic solution for each $s > s_0$.

Finally, let u be a bounded solution of (102) for some $s < s_0$. The equation easily implies that u' is also bounded and given $u_+ > \sup u(t) \geq \inf u(t) > u_-$ with $|u_\pm|$ large enough, let us consider the associated truncated equation (107). By construction, u is also a solution of this equation (for which the Cauchy problem is globally uniquely solvable) with $|u| + |u'|$ bounded. A theorem of MASSERA [80] implies then that (107) should have a T-periodic solution, a contradiction with Lemma 7.3 and the nonexistence of T-periodic solution for (102) when $s < s_0$. Thus, for those values of s, (102) has no bounded solution. ■

Remark 7.4. Theorem 7.3 can be considered as a partial extension to a second order differential equation of the result of [92] which gives a complete qualitative picture of the set of solutions for first order scalar equations with convex and coercive nonlinearities.

Remark 7.5. More results on equation (102), including the existence of some subharmonics of order 2 can be found in [112].

Remark 7.6. Theorems 7.1 and 7.2 also provide information about the number and the stability of the T-periodic solutions of second order forced dissipative equations of pendulum-type. See [111] and [112] for details.

References

[1] V.V. Amel'kin, I.V. Gaishun and N.N. Ladis, Periodic solutions in the case of a constantly acting perturbation, Differential equations, 11 (1975), 1569-1573

[2] A. Bahri and H. Berestycki, Existence of forced oscillations for some nonlinear differential equations, Comm. Pure Appl. Math. 37 (1984), 403-442

[3] A. Bahri and H. Berestycki, Forced vibrations of superquadratic Hamiltonian systems, Acta Math. 152 (1984), 143-197

[4] Th. Bartsch and J. Mawhin, The Leray-Schauder degree of S^1-equivariant operators associated to autonomous neutral equations in spaces of periodic functions, J. Differential Equations, 92 (1991), 90-99

[5] J.W. Bebernes and K. Schmitt, Periodic boundary value problems for systems of second order differential equations, J. Differential Equations, 13 (1973), 32-47

[6] I. Berstein and A. Halanay, The index of a critical point and the existence of periodic solutions to a system with small parameter, Doklady Ak. Nauk. 111 (1956), 923-925 (translation and remarks by G. B. Gustafson)

[7] N.A. Bobylev, The construction of regular guiding functions, Soviet Math. Dokl., 9 (1968), 1353-1355

[8] L.E.J. Brouwer, Ueber Abbildungen von Mannigfaltigkeiten, Math. Ann. 71 (1912), 97-115

[9] A. Capietto. Continuation theorems for periodic boundary value problems, Ph.D. Thesis, SISSA, Trieste, 1990

[10] A. Capietto, J. Mawhin, and F. Zanolin, Continuation theorems for periodic perturbations of autonomous systems, Trans. Amer. Math. Soc., to appear.

[11] A. Capietto, J. Mawhin and F. Zanolin, The coincidence degree of some functional differential operators in spaces of periodic functions and related continuation theorems, in Delay Differential Equations and Dynamical Systems (Claremont 1990), Busenberg, Martelli eds, Springer, Berlin, 1991, 76-87.

[12] A. Capietto, J. Mawhin and F. Zanolin, A continuation approach to superlinear periodic boundary value problems, J. Differential Equations 88 (1990), 347-395

[13] A. Capietto, J. Mawhin and F. Zanolin, Periodic solutions of some superlinear functional differential equations, in Ordinary and functional Differential Equations, Kyoto 1990, World Scientific, Singapore, 1991, 19-31

[14] A. Capietto, J. Mawhin and F. Zanolin, A continuation approach to superlinear two point boundary value problems, in preparation

[15] A. Capietto and F. Zanolin, An existence theorem for periodic solutions in convex sets with applications, Results in Mathematics, 14 (1988), 10-29

[16] A. Capietto and F. Zanolin, A continuation theorem for the periodic BVP in flow-invariant ENRs with applications, J. Differential Equations, 83 (1990), 244-276

[17] J. Cronin, The number of periodic solutions of nonautonomous systems, Duke Math. J., 27 (1960). 183-194

[18] J. Cronin, Lyapunov stability and periodic solutions, Bol. Soc. Mat. Mexicana, (1965), 22-27

[19] J. Cronin, The point at infinity and periodic solutions, J. Differential Equations, 1 (1965), 156-170

[20] J. Cronin, Periodic solutions of some nonlinear differential equations, J. Differential Equations, 3 (1967), 31-46

[21] J. Cronin, Periodic solutions of nonautonomous equations, Boll. Un. Mat. Ital., (4) 6 (1972), 45-54

[22] E.N. Dancer, Boundary value problems for weakly nonlinear ordinary differential equations, Bull. Austral. Math. Soc., 15 (1976), 321-328

[23] E.N. Dancer, Symmetries, degree, homotopy indices and asymptotically homogeneous problems, Nonlinear Analysis, TMA, 6 (1982), 667-686

[24] K. Deimling, Nonlinear functional analysis, Springer, Berlin, 1985

[25] L. Derwidué, Systèmes différentiels non linéaires ayant des solutions périodiques, Acad. Royale de Belgique, Bull. Cl. des Sciences (5) 49 (1963), 11-32, 82-90; (5) 50 (1964), 928-942, 1130-1142

[26] L. Derwidué, Etude géométrique d'une équation différentielle non linéaire, Bull. Soc. Royale Sciences Liège 34 (1965), 180-187

[27] L. Derwidué, Systèmes différentiels non linéaires ayant des solutions uniformément bornées dans le futur, Bull. Soc. Royale Sciences Liège 34 (1965), 555-572

[28] L. Derwidué, L'équation de Forbat, in Colloque du CBRM sur les équations différentielles non linéaires, leur stabilité et leur périodicité, vander, Louvain, 1970, 7-27

[29] T. Ding, R. Iannacci and F. Zanolin, On periodic solutions of sublinear Duffing equation, J. Math. Anal. Appl. 158 (1991), 316-332

[30] W.Y. Ding, Generalizations of the Borsuk theorem, J. Math. Anal. Appl. 110 (1985) 553-567

[31] C.L. Dolph, Nonlinear integral equations of the Hammerstein type, Trans. Amer. Math. Soc. 66 (1949), 289 307

[32] H. Ehrmann. Ueber die Existenz der Lösungen von Randwertaufgaben bei gewöhnlichen nichtlinearen Differentialgleichungen zweiter Ordnung, Math. Ann. 134 (1957), 167-194

[33] C. Fabry, J. Mawhin and M. Nkashama, A multiplicity result for periodic solutions of forced nonlinear second order differential equations, Bull. London Math. Soc. 18 (1986), 173-180

[34] R. Faure, Solutions périodiques d'équations différentielles et méthode de Leray-Schauder (Cas des vibrations forcées), Ann. Inst. Fourier (Grenoble) 14 (1964), 195-204

[35] R. Faure, Sur l'application d'un théorème de point fixe à l'existence de solutions périodiques, C.R. Acad. Sci. Paris, 282 (1976), A, 1295-1298

[36] R. Faure, Solutions périodiques d'équations admettant des pôles : étude par la méthode de Leray Schauder et par un théorème de point fixe, C.R. Acad. Sci. Paris, 283 (1976), A, 481-484

[37] L. Fernandes and F. Zanolin, Periodic solutions of a second order differential equation with one-sided growth restrictions on the restoring term, Arch. Math. (Basel), 51 (1988), 151-163

[38] A. Fonda and P. Habets, Periodic solutions of asymptotically positively homogeneous differential equations, J. Differential Equations, 81 (1989), 68-97

[39] A. Fonda and F. Zanolin, Periodic solutions to second order differential equations of Liénard type with jumping nonlinearities, Comment. Math. Univ. Carolin., 28 (1987), 33-41

[40] N. Forbat and A. Huaux, Détermination approchée et stabilité locale de la solution périodique d'une équation différentielle non linéaire, Mém. et Public. Soc. Sciences, Artts Lettres du Hainaut 76 (1962), 3-13

[41] S. Fučik, Solvability of nonlinear equations and boundary value problems, D. Reidel Publishing Company, Dordrecht, 1980.

[42] S. Fučik and V. Lovicar, Periodic solutions of the equation $x'' + g(x) = p$, Časopis Pest. Mat., 100 (1975), 160-175

[43] M. Furi, M. Martelli and A. Vignoli, On the solvability of nonlinear operator equations in normed spaces, Ann. Mat. Pura Appl. (4) 124 (1980), 321-343

[44] M. Furi and M.P. Pera, An elementary approach to boundary value problems at resonance, J. Nonlinear Anal. 4 (1980) 1081-1089

[45] M. Furi and M.P. Pera, Co-bifurcating branches of solutions for nonlinear eigenvalue problems in Banach spaces, Annali Mat. Pura Appl. 135 (1983), 122

[46] M. Furi and M. P. Pera, A continuation principle for forced oscillations on differentiable manifolds, Pacific J. Math., 121 (1986), 321-338.

[47] M. Furi and M.P. Pera, The forced spherical pendulum does have forced oscillations, in Delay Differential Equations and Dynamical Systems (Claremont 1990), Busenberg and Martelli eds, Springer, Berlin, 1991, 176-182

[48] R.E. Gaines and J. Mawhin, Coincidence degree and nonlinear differential equations, Lecture Notes in Math., 586, Springer-Verlag, Berlin, 1977

[49] K. Geba, A. Granas, T. Kaczynski and W. Krawcewicz, Homotopie et équations non linéaires dans les espaces de Banach, C.R. Acad. Sci. Paris 300, ser. I (1985), 303-306

[50] R.E. Gomory. Critical points at infinity and forced oscillations, in Contributions to the theory of nonlinear oscillations, 3, Ann. of Math. Studies, 36, Princeton Univ. press, N. J. 1956, pp. 85-126

[51] W.B. Gordon, Conservative dynamical systems involving strong forces, Trans. Amer, Math. Soc. 204 (1975), 113-135

[52] W.B. Gordon, A minimizing property of Keplerian orbits, Amer. J. Math. 99 (1977), 961-971

[53] A. Granas, The Leray-Schauder index and the fixed point theory for arbitrary ANRs, Bull. Soc. Math. France, 100 (1972), 209-228

[54] P. Habets and L. Sanchez, Periodic solutions of dissipative dynamical systems with singular potentials, J. Differential and Integral Equations, 3 (1990), 1139-1149

[55] P. Habets and L. Sanchez, Periodic solutions of some Liénard equations with singularities, Proc. Amer. Math. Soc.109 (1990), 1035-1044

[56] A. Halanay, În legătură cu metoda parametrolui mic (Relativement à la méthode du petit paramètre), Acad. R. P. R., Bul. St., Sect. Mat. fiz., 6 (1954), 483-488

[57] A. Halanay, Solutions périodiques des systèmes non-linéaires à petit paramètre, Rend. Accad. Naz. Lincei (Cl. Sci. Fis. Mat. Natur.), 22 (Ser. 8) (1957), 30-32

[58] A. Halanay, Differential Equations, stability, oscillations, time lags, Academic Press, New-York, London, 1966

[59] J.K. Hale and A.S. Somolinos, Competition for fluctuating nutrient, J. Math. Biology, 18 (1983), 255-280

[60] A. Huaux, Sur l'existence d'une solution périodique de l'équation différentielle non linéaire $x'' + 0.2x' + x/(1-x) = (0,5)\cos \omega t$, Bull. Cl. Sciences Acad. R. Belgique (5) 48 (1962), 494-504

[61] R. Iannacci and M. Martelli, Branches of solutions of nonlinear operator equations in the atypical bifurcation case, preprint.

[62] Yu.S. Kolesov, Study of stability of solutions of second-order parabolic equations in the critical case, Math. USSR-Izvestija 3 (1969), 1277-1291

[63] M.A. Krasnosel'skii, The operator of translation along the trajectories of differential equations, Amer. Math. Soc., Providence, R.I., 1968

[64] M.A. Krasnosel'skii and P.P. Zabreiko, Geometrical methods of nonlinear Analysis, Springer-Verlag, Berlin, 1984

[65] L. Kronecker, Ueber Systeme von Funktionen mehrerer Variabeln, Monatsber. Berlin Akad. (1869), 159-193

[66] B. Laloux and J. Mawhin, Coincidence index and multiplicity, Trans. Amer. Math. Soc. 217 (1976), 143-162

[67] A. Lando, Periodic solutions of nonlinear systems with forcing term, J. Differential Equations, 3 (1971), 262-279

[68] A. Lando, Forced oscillations of two-dimensional nonlinear systems, Applicable Analysis, 28 (1988), 285-295

[69] A. Lasota, Une généralisation du premier théorème de Fredholm et ses applications à la théorie des équations différentielles ordinaires, Ann. Polon. Math., 18 (1966), 65-77

[70] A. Lasota and Z. Opial, Sur les solutions périodiques des équations différentielles ordinaires, Ann. Polon. Math., 16 (1964), 69-94

[71] A.C. Lazer and P.J. McKenna, A semi-Fredholm principle for periodically forced systems with homogeneous nonlinearities, Proc. Amer. Math. Soc., 106 (1989), 119-125

[72] A.C. Lazer and P.J. McKenna, On the existence of stable periodic solutions of differential equations of Duffing type, Proc. Amer. Math. Soc. 110 (1990), 125-133

[73] A.C. Lazer and P.J. McKenna, Nonresonance conditions for the existence, uniqueness and stability of periodic solutions of differential equations with a symmetric nonlinearity, Differential and Integral Equations 4 (1991), 719-730

[74] A.C. Lazer and S. Solimini, On periodic solutions of nonlinear differential equations with singularities, Proc. Amer. Math. Soc. 99 (1987), 109-114

[75] J. Leray, Les problèmes non linéaires, L'enseignement math. 35 (1936), 139-151

[76] J. Leray and J. Schauder, Topologie et equations fonctionelles, Ann. Sci. Ecole Norm. Sup. (3) 51 (1934), 45-78

[77] N.G. Lloyd, Degree theory, Cambridge Univ. Press, Cambridge, 1978

[78] Y. Long, Multiple solutions of perturbed superquadratic second order Hamiltonian systems, Trans. Amer. Math. Soc. 311 (1989), 749-780

[79] W.S. Loud, Periodic solutions of $x'' + cx' + g(x) = \varepsilon f(t)$, Mem. Amer. Math. Soc., 31 (1959)

[80] J.L. Massera. The existence of periodic solutions of systems of differential equations, Duke Math. J. 17 (1950), 457-475

[81] J. Mawhin, Equations intégrales et solutions périodiques des systèmes différentiels non linéaires. Acad. Roy. Belg. Bull. Cl. Sci., 55 (1969), 934-947

[82] J. Mawhin, Existence of periodic solutions for higher order differential systems that are not of class D, J. Differential Equations, 8 (1970), 523-530

[83] J. Mawhin, Equations fonctionnelles non linéaires et solutions périodiques, in Equadiff 70, Centre de Recherches Physiques, Marseille, 1970

[84] J. Mawhin, Periodic solutions of nonlinear functional differential equations, J. Differential Equations, 10 (1971), 240-261

[85] J. Mawhin, Equivalence theorems for nonlinear operator equations and coincidence degree theory for some mappings in locally convex topological vector spaces, J. Differential Equations 12 (1972), 610-636

[86] J. Mawhin, Nonlinear perturbations of Fredholm mappings in normed spaces and applications to differential equations, Trabalho de Mat. 61, Univ. de Brasilia, 1974

[87] J. Mawhin, Periodic solutions of some perturbed differential systems, Boll. Un. Mat. Ital., 11 (4) (1975), 299-305

[88] J. Mawhin, Periodic solutions, Workshop on nonlinear boundary value problems for ordinary differential equations and applications, Trieste, 1977, ICTP SMR/42/1

[89] J. Mawhin, Topological degree methods in nonlinear boundary value problems, CBMS 40, Amer. Math. Soc., Providence, R.I., 1979

[90] J. Mawhin, Compacité, monotonie et convexité dans l'étude des problèmes aux limites semi linéaires, Sémin. Anal. Moderne, 19, Univ. Sherbrooke, 1981

[91] J. Mawhin, Point fixes, point critiques et problemes aux limites, Séminaire de Mathématiques Supérieures, vol. 92, Les Presses de l'Université de Montréal, 1985

[92] J. Mawhin, Qualitative behavior of the solutions of periodic first order scalar differential equations with strictly convex coercive nonlinearity, in Dynamics of infinite dimensional systems, Chow and Hale eds., Springer, Berlin, 1987, 151-159

[93] J. Mawhin, A simple proof of a semi-Fredholm principle for periodically forced systems with homogeneous nonlinearities, Arch. Math. (Brno), 25 (1989), 235-238

[94] J. Mawhin, Bifurcation from infinity and nonlinear boundary value problems, in Ordinary and Partial Differential equations, (Dundee 1989), Pitman, 1990, 119-129

[95] J. Mawhin and C. Munoz, Application du degré topologique à l'estimation du nombre des solutions périodiques d'équations différentielles, Annali Mat. Pura Appl. (4) 96 (1973), 1-19

[96] J. Mawhin and C. Rybakowski, Continuation theorems for semi-linear equations in Banach spaces : a survey, in Nonlinear Analysis, World Scientific, Singapore, 1987, 367-405

[97] J. Mawhin and K. Schmitt, Landesman-Lazer type problems at an eigenvalue of odd multiplicity. Results in Math. 14 (1988), 138-146

[98] J. Mawhin and K. Schmitt, Nonlinear eigenvalue problems with the parameter near resonance. Ann. Polon. Math. 60 (1990), 241-248

[99] J. Mawhin and J.R. Ward, Periodic solutions of some forced Liénard differential equations at resonance, Arch. Math. (Basel), 41 (1983), 337-351

[100] J. Mawhin and M. Willem, Critical Point Theory and Hamiltonian Systems, Appl. Math. Sci., vol. 74, Springer, New York, 1989

[101] G.R. Morris, An infinite class of periodic solutions of $x'' + 2x^3 = p(t)$, Proc. Cambridge Phil. Soc. 61 (1965), 157-164

[102] E. Muhamadiev, Construction of a correct guiding function for a system of differential equations, Soviet Math. Dokl., 11 (1970), 202-205

[103] E. Muhamadiev, On the theory of periodic solutions of systems of ordinary differential equations, Soviet Math. Dokl., 11 (1970), 1236-1239

[104] W. Müller, Ueber die Beschränkheit der Lösungen der Cartwright-Littlewoddschen Gleichung, Math. Nachr. 29 (1965), 25-40

[105] W. Müller, Qualitative Untersuchungen der Lösungen nichtlinearer Differentialgleichungen zweiter Ordnung nach der direkten Methode von Liapounov, Abhandlungen Deutsche Akad. Wiss., Kl. Math. Phys. Techn., Heft 4, Berlin, 1965

[106] L. Nirenberg, Comments on nonlinear problems, Le Matematiche (Catania) 36 (1981) 109-119

[107] R. D. Nussbaum, The fixed point index and some applications, Séminaire de Mathématiques Supérieures, vol. 94, Les Presses de l'Université de Montréal, 1987

[108] P. Omari, Gab. Villari and F. Zanolin, Periodic solutions of the Liénard equation with one-sided growth restrictions, J. Differential Equations, 67 (1987), 278-293

[109] P. Omari and F. Zanolin, On forced nonlinear oscillations in n-th order differential systems with geometric conditions, Nonlinear Analysis, TMA, 8 (1984), 723-784

[110] Z. Opial, Sur les solutions périodiques de l'équation différentielle $x'' + g(x) = p(t)$, Bull. Acad. Polon. Sci. Sér. Sci. Math. Astr. Phys., 8 (1960), 151-156

[111] R. Ortega, Stability and index of periodic solutions of an equation of Duffing type, Boll. Un. Mat. Ital. (7) 3-B (1989), 533-546

[112] R. Ortega, Stability of a periodic problem of Ambrosetti-Prodi type, Differential and Integral Equations, 3 (1990), 275-284

[113] R. Ortega, Topological degree and stability of periodic solutions for certain differential equations, J. London Math. Soc. (2) 42 (1990), 505-516

[114] R. Ortega, A criterion for asymptotic stability based on topological degree, to appear

[115] R. Ortega, The first interval of stability of a periodic equation of Duffing type, Proc. Amer. Math. Soc., to appear

[116] J. Pejsachowicz and A. Vignoli, On the topological coincidence degree for perturbations of Fredholm operators, Boll. Un. Mat. Ital. (5) 17-B (1980), 1457-1466

[117] V.A. Pliss, Nonlocal problems of the theory of oscillations, Academic Press, New York, 1966

[118] P.J. Rabier, Topological degree and the theorem of Borsuk for general covariant mappings with applications, Nonlinear Analysis 16 (1991) 399-420

[119] P. Rabinowitz. On bifurcation from infinity, J. Differential Equations 14 (1973), 462-475

[120] P. Rabinowitz, Multiple critical points of perturbed symmetric functionals, Trans. Amer. Math. Soc. 272 (1982), 753-769

[121] R. Reissig, Periodic solutions of a nonlinear n-th order vector differential equation, Ann. Mat. Pura Appl., (4) 87 (1970), 111-123

[122] R. Reissig, G. Sansone and R. Conti, Qualitative Theorie nichtlinearer Differentialgleichungen, Cremonese, Roma, 1963

[123] R. Reissig, G. Sansone and R. Conti, Non-linear Differential Equations of Higher Order, Noordhoff, Leyden, 1974

[124] N. Rouche and J. Mawhin, Ordinary differential equations, Pitman, London, 1980

[125] J. Schauder, Der Fixpunktsatz in Funktionalraümen Studia Math. 2 (1930), 171-180

[126] K. Schmitt, Periodic solutions of small period of systems of n-th order differential equations, Proc. Amer. Math. Soc., 36 (1972), 459-463

[127] R.A. Smith, Massera's convergence theorem for periodic nonlinear differential equations, J. Math. Anal. Appl. 120 (1986), 679-708

[128] S. Solimini, On forced dynamical systems with a singularity of repulsive type, J. Nonlinear Analysis 14 (1990), 489-500

[129] R. Srzednicki. On rest points of dynamical systems, Fund. Math., 126 (1985), 69-81

[130] R. Srzednicki. Periodic and constant solutions via topological principle of Ważewski, Acta Math. Univ. Iag., 26 (1987), 183-190

[131] H. Steinlein. Borsuk's antipodal theorem and its generalizations and applications : a survey, in Méth. Topologiques en Analyse Non-Linéaire, Granas ed., Sémin. Math. Sup. n 95, Montréal, 1985, 166-235

[132] F. Stoppelli. Su un'equazione differenziale della meccanica dei fili, Rend. Accad. Sci. Fis. Mat. Napoli, 19 (1952), 109-114

[133] M. Struwe, Multiple solutions of anticoercive boundary value problems for a class of ordinary differential equations of the second order, J. Differential Equations 37 (1980), 285-295

[134] M. Urabe, Nonlinear autonomous oscillations. Analytical theory, Academic Press, New York, 1967

[135] G. Villari, Soluzioni periodiche di una classe di equazioni differenziali, Ann. Mat. Pura Appl. (4) 73 (1966), 103-110

[136] P. Volkmann, Zur Definition des Koinzidenzgrades, preprint, 1981

[137] P. Volkmann, Démonstration d'un théorème de coincidence par la méthode de Granas, Bull. Soc. Math. Belgique, B 36 (1984), 235-242

[138] Z.Q. Wang : Symmetries and the calculations of degree, Chinese Ann. of Math. 108 (1989) 520-536

[139] J.R. Ward, Asymptotic conditions for periodic solutions of ordinary differential equations, Proc. Amer. Math. Soc. 81 (1981), 415-420

[140] J.R. Ward, Conley index and non-autonomous ordinary differential equations, Results in Mathematics, 14 (1988), 191-209

[141] E. Zeidler, Nonlinear functional analysis and its applications, vol. I, Springer, New York, 1986

The Fixed Point Index and Fixed Point Theorems

Roger D. Nussbaum

Mathematics Department, Hill Center
Rutgers University
New Brunswick, New Jersey 08903

Introduction: If X is a "nice" topological space W is an open subset of X, and $f: W \to X$ is a continuous map which has a compact (possibly empty) fixed point set $S = \{x \in W: f(x) = x\}$ and which is "nice" on a neighborhood of S, then one can define an integer $i_X(f,W)$, called the fixed point index of f on W. If $W = X$, $i_X(f,X)$ is the famous Lefschetz number of $f: X \to X$; and if X is a Banach space, and f is compact on a neighborhood of S, $i_X(f,W) = \deg(I - f, W, 0)$, the Leray–Schauder degree of $I - f$ on W. The fixed point index provides a framework for studying fixed point theory in a generality useful to the analyst. For example, we shall later give results which unify and generalize the Schauder fixed point theorem, the Lefschetz–Hopf fixed point theorem, and Darbo's fixed point theorem. If X is a closed, convex subset of a Banach space E or a "metric ANR" in E, the fixed point index enables one to mimic various degree theory arguments, even though X may have empty interior in E. Thus we shall later describe a global bifurcation theorem for a map $f: X \times (0,\infty) \to X$, where X is a closed, convex subset of a Banach space.

As we have tried to suggest, the fixed point index is a very useful tool for the nonlinear analyst and indeed, has even had useful applications in the study of positive linear operators (see [52] and [53]). Nevertheless, in comparison to Leray–Schauder degree, the fixed point index is not well–known.

In these notes we begin in Section 1 by describing the "classical" fixed point index, i.e., the case in which X is a compact metric ANR (for example, a finite union of compact, convex sets in a Banach or a retract of such a space). We indicate how the Leray–Schauder degree can be derived from the classical fixed point index, and we show how the fixed point index can be generalized to the case that X is a metric ANR and f is compact on some open neighborhood of S.

The fixed point index provides a powerful tool for proving fixed point theorems. For example, the famous Lefschetz–Hopf fixed point theorem is essentially built into the framework of the fixed point index. In Section 2 we describe a variety of results which are either explicit or implicit in [43] and [49] but which are surprisingly little–known. We call the reader's attention to Corollary 2 and Theorem 12, which are implicit in [49] and which contain A. Tromba's results in [62] as a very special case. (We believe the significance of Tromba's work in [62] lies in the novelty of the approach rather than the novelty of the results.) In the course of proving the various fixed point theorems in Section 2, we are essentially forced to consider the problem of generalizing the fixed point index. Theorem 11 describes various classes of maps for which this generalized fixed point index is defined. Other examples, notably maps f of the form gr, where g is a k–set–contraction with $k<1$ and r is C^1 retraction of an open set U in a Banach space E onto a Banach manifold $M \subset U$, are discussed in [49].

Much of the abstract motivation for our results comes from an old conjecture. If G is a closed, bounded convex subset of a Banach space and $f: G \to G$ is a continuous map such that $\overline{f^N(G)}$ is compact for some $N>1$, does f necessarily have a fixed point? In general, the answer is unknown, although the results of Section 2 show that, under mild further assumptions, the answer is "yes".

There are many applications of the fixed point index. Section 3 discusses an application which happens to be of interest to the author, namely the existence of periodic solutions of the nonlinear differential–delay equation

$$x'(t) = -\lambda g(x(t-1)), \quad \lambda>0.$$

We assume that g is continuous, that $ug(u) > 0$ for $u \neq 0$ and that $\lim_{u\to 0^+} g(u)u^{-1} = a$ and $\lim_{u\to 0^-} g(u)u^{-1} = b$. The general approach to such problems is familiar, but because $a \neq b$, our results are new and the argument requires several twists, eg, the use of a non–standard global bifurcation theorem (see Theorem 3 and Corollary 3). Statements about periodic solutions of the differential–delay equation can be found in Theorem 4, Corollary 4, Theorem 5 and Remark 4 of Section 3.

In the course of proving Theorems 4 and 5, we observe (see Theorem 1) that if $f: W \subset X \to X$ is a continuous map, $x_0 \in W$ is an "asymptotically stable" fixed point of f, U is a small open neighborhood of x_0 and the fixed point index of $f: U \to X$ is defined, then $i_X(f,U) = 1$. Even if f is C^1 at x_0, this is a topological result which does not follow from linearization at x_0. Although Theorem 1 is a generalization of old results, it does not seem well-known even when f is compact.

1. The classical fixed point index.

Suppose that X is a topological space, W is an open subset of X and $f: W \to X$ is a continuous map such that $S: = \{x \in W | f(x) = x\}$ is compact or empty. Under further assumptions on f and X, one wants to associate in this situation an integer $i_X(f,W)$ (sometimes written $i(f,W,X)$), called the fixed point index of $f: W \to X$. Roughly speaking, $i_X(f,W)$ will be an algebraic count of the number of fixed points of f in W. If X is a Banach space or normed linear space and $\overline{f(W)}$ is compact, one wants $i_X(f,W)$ to be related to the Leray–Schauder degree of $I-f$ in thc natural way: $i_X(f,W) = \deg(I-f,W,0) =$ the Leray–Schauder degree of $I-f$ on W with respect to 0. Thus the fixed point index generalizes the Leray–Schauder degree. However, it is also defined in situations where degree theory is not directly applicable, for example, when X is a closed convex set with empty interior in a Banach space Y. As we shall see, the fixed point index also generalizes the famous Lefschetz fixed point theorem. Because of these connections, the fixed point index is a powerful, though surprisingly little–known, tool for the study of nonlinear problems in analysis. Indeed the fixed point index is even useful in the study of <u>linear</u> positive operators: See Section 2 of [52] and [53].

In this section we shall take an axiomatic approach. We shall describe a framework within which the fixed point index can be defined, and we shall give four properties of the fixed point index which, in fact, determine it uniquely. We shall present few details of the development of the "classical" fixed point index and refer the reader to [7], [8], [10], [12], [13], [14], [16], [23], [32], [33], [34] and [60] for background. Here we take the viewpoint that the fixed point index is a tool which the analyst can use if he or she knows a few basic facts.

Once we have described the "classical fixed point index," we shall end this section and continue in the next section by showing how the fixed point index can be extended to more general classes of maps f and more general spaces X.

Again we shall take the view that it is the properties of the extension rather than the details of the construction and definition which are important.

The first difficulty in defining the fixed point index is that one must find a suitable class of spaces X. If X is compact, $H_*(X)$ denotes Čech homology with coefficients in the rationals, and $H_*(X)$ is a finite dimensional graded vector space, one wants

$$i_X(f,X) = L(f,X) := \sum_{i\geq 0} (-1)^i \text{ trace } (f_{*,i}),$$

where $L(f,X)$ denotes the Lefschetz number of $f: X \to X$ and $f_{*,i}$: $H_i(X) \to H_i(X)$. It is known that the Lefschetz number in this case, although, a priori, only a rational number, is actually an integer. One would like to know that if $L(f,X) \neq 0$, then f has a fixed point; and, indeed, if X is a compact simplicial complex and $L(f,X) \neq 0$, the classical Lefschetz–Hopf theorem asserts that f has a fixed point in X. However, for general compact spaces X, the Lefschetz fixed point theorem is false. There exists a compact space X, with X contractible in itself to a point, and a continuous map $f: X \to X$ which has no fixed point. In this situation, f is homotopic to a point map, so $L(f,X) = 1$. We refer the reader to [2] for further details.

Because of the above remarks, it seems necessary to restrict the class of spaces X and/or the class of maps f for which one defines a fixed point index. At the very least, in view of applications in analysis, one needs to include compact sets X in a Banach space E such that $X = \bigcup_{i=1}^{m} C_i$, where $C_i \subset E$ is a compact, convex set for $1 \leq i \leq m$. A reasonable class of spaces X in which to try to study the fixed point index is provided by the so–called metric absolute neighborhood retracts or metric ANR's. Recall that a metrizable space X is called a metric ANR if whenever A is a closed subspace of a metric space M and $f: A \to X$ is a continuous map, there exists an open neighborhood U of A in M and a continuous map $F: U \to X$ with $F(x) = f(x)$ for all $x \in A$. X is called a metric absolute retract or metric AR if, in the above situation, there always exists a map $F: M \to X$ with $F(x) = f(x)$ for all $x \in A$. Recall that if X_1 is a closed subset of a topological space Y, X_1 is called a "retract of Y" if there exists a continuous map $r: Y \to X_1$ such that $r(y) \in X_1$ for all $y \in Y$ and $r(x) = x$ for all $x \in X_1$. Such a map r is called a retraction of Y onto X_1. If X is a metric AR, M is a metric space, $X_1 \subset M$ is a closed subset of M, and $h: X \to X_1$ is a homeomorphism, then one can derive from the definition of an AR that there exists a retraction $r: M \to X_1$ of M onto X_1. If X is a metric ANR, there exists an open neighborhood U of X_1 and a retraction $r: U \to X_1$.

We now recall some basic facts about metric ANR's.

(1) (See J. Dugundji [15]). Suppose that X is a convex subset of a locally convex topological vector space E, A is a closed subset of a metric space M, and $f: A \to X$ is a continuous map. Then f has a continuous extension $F: M \to X$. In particular, if X is a closed, convex subset of a metrizable locally convex topological vector space E (or lctvs E, for short), there exists a retraction $r: E \to X$. (Take $X = A$, $M = E$ and $f(x) = x$ for all $x \in A$).

(2) (See K. Borsuk [3]). Suppose that X is a metric space and that X_0 and X_1 are subspaces of X such that X_0, X_1 and $X_0 \cap X_1$ are metric ANR's. Then it follows that $X_0 \cup X_1$ is a metric ANR.

(3) Suppose that C_i, $1 \le i \le m$, are convex subsets of an lctvs E and that $X = \bigcup_{i=1}^{m} C_i$ is metrizable. Then X is a metric ANR. If $m=1$, this is simply property 1 above. To obtain the general case, use (1) and (2) and induction on m. In particular, it follows that any compact simplicial complex is an ANR. Also, if $C_1, C_2, \cdots, C_m$ are closed, convex subsets of a metrizable lctvs E and $X = \bigcup_{i=1}^{m} C_i$, there exists an open neighborhood U of X in E and a continuous retraction $r: U \to X$.

(4) (See O. Hanner, [25] and [26]). Any open subset U of a metric ANR X is itself a metric ANR. Conversely, suppose that X is a metric space and that, for each $x \in X$, there exists an open neighborhood U_x of x in X such that U_x is a metric ANR. Then X is a metric ANR. In particular, any metrizable Banach manifold is an ANR.

(5) If X is a metric ANR, $Y \subset X$, and there exists a retraction r of X onto Y, then Y is a metric ANR.

With these preliminaries, we can describe the classical fixed point index more precisely. The reader can find more detail in the previously cited references.

Suppose that X is a compact metric ANR, W is an open subset of X (possibly empty) and $f: W \to X$ is a continuous map such that $S = \{x \in W \mid f(x) = x\}$ is compact (possibly empty). Then there is defined an integer $i_X(f,W)$, called the fixed point index of $f: W \to X$. It is frequently assumed that f is defined and continuous on $\overline{W}$ and that $f(x) \neq x$ for $x \in \overline{W} - W$, and these assumptions certainly imply that $S = \{x \in W \mid f(x) = x\}$ is compact. However, the direct assumption that S is compact is more general and provides greater flexibility in applications. Furthermore, if one assumes S is compact and one takes V to be an open neighborhood of S such that $\overline{V} \subset W$, one can prove that $i_X(f,V)$ is independent of V and one can define $i_X(f,W) = i_X(f,V)$ for V as above.

If (f,W,X) is as above, the map $(f,W,X) \to i_X(f,W)$ satisfies several properties which are usually called the additivity, homotopy, normalization and commutativity properties. Together these properties uniquely determine the fixed point index, at least for continuous maps defined in compact metric

1. (Additivity). Suppose that X is a compact metric ANR. Let W be an open subset of X and $f: W \to X$ a continuous map. If f has no fixed points in W, $i_X(f,W) = 0$. If W_1 and W_2 are open subsets of W, $S_j = \{x \in W_j \mid f(x) = x\}$ is a compact (possibly empty) subset of W_j for $j = 1,2$, and $S_1 \cap S_2$ is empty, then

$$i_X(f,W_1 \cup W_2) = i_X(f,W_1) + i_X(f,W_2).$$

If X is a compact metric ANR, W is an open subset of X and $f: W \to X$ is a continuous map for which $S = \{x \in W \mid f(x) = x\}$ is compact, the additivity property implies that if V is any open neighborhood of S with $V \subset W$, then $i_X(f,V) = i_X(f,W)$.

2. (Homotopy). Let X be a compact metric ANR. Assume that Ω is an open subset of $X \times [0,1]$ and that $f: \Omega \to X$ is a continuous map such that $\Sigma = \{(x,t) \in \Omega \mid f(x,t) = x\}$ is compact. For $0 \leq t \leq 1$, let $\Omega_t = \{x \mid (x,t) \in \Omega\}$ and define $f_t: \Omega_t \to X$ by $f_t(x) := f(x,t)$. Then $i_X(f_t,\Omega_t)$ is constant for $0 \leq t \leq 1$. (If Ω_t is empty, we define $i_X(f_t,\Omega_t) = 0$.)

3. (Commutativity). Suppose that X and Y are compact, metric ANR's, that U and V are open subsets of X and Y respectively and that $f: U \to Y$ and $g: V \to X$ are continuous maps. Thus we have that $gf: f^{-1}(V) \to X$ and $fg: g^{-1}(U) \to Y$. Assume that

$S = \{x \in f^{-1}(V)\colon g(f(x)) = x\}$ is compact (possibly empty). Then it follows that $T = \{y \in g^{-1}(U)\colon g(f(y)) = y\}$ is compact and

$$i_X(gf, f^{-1}(V)) = i_Y(fg, g^{-1}(U)).$$

An important special case of the commutativity property occurs when $Y \subset X$ and Y inherits its topology from X. Suppose that U is an open subset of X, $f\colon U \to X$ is a continuous map and $S = \{x \in U \mid f(x) = x\}$ is compact. Assume also that $f(U) \subset Y$, so we can consider f as a map from U to Y. We can also consider f as a map from U to X, and to distinguish $f\colon U \to Y$ from $f\colon U \to X$ we write $\tilde{f}$ for the latter map. If $i\colon Y \to X$ is the inclusion map, $\tilde{f} = if$. The commutativity property implies (taking $V = Y$, $g = i$) that

$$i_X(if,\ U) = i_Y(fi,\ U \cap Y).$$

Usually, this is written in the less precise form

$$i_X(f,U) = i_Y(f, U \cap Y),$$

which is valid when $f(U) \subset Y \subset X$.

If X is a compact metric ANR, it is known that $H_*(X)$ is independent of the particular homology theory. Thus we shall think of $H_*(X)$ as being singular homology with rational coefficients. It is also known that $H_i(X)$ is a finite dimensional vector space over the rationals and that $H_i(X) = \{0\}$ for all sufficiently large i. Thus $L(f,X)$, the Lefschetz number of $f\colon X \to X$, is defined:

$$L(f,X) = \sum_{i \geq 0} (-1)^i \ \text{trace} \ (f_{*,i}).$$

Here, trace $(f_{*,i})$ of course denotes the trace of the vector space endomorphism $f_{*i}\colon H_i(X) \to H_i(X)$.

A. (Normalization). Let X be a compact metric ANR and $f\colon X \to X$ a continuous map. Then we have

$$i_X(f,X) = L(f,X) = \sum_{i \geq 0} (-1)^i \ \text{trace} \ (f_{*i}),$$

so f has a fixed point in X if $L(f,X) \neq 0$.

The analyst will note that the fixed point index has not even been defined in the generality of the Leray–Schauder degree, i.e., for the case that X is a Banach space, W is a bounded open subset of X, and $f: \overline{W} \longrightarrow X$ is a continuous map with $\overline{f(\overline{W})}$ compact and $f(x) \neq x$ for all $x \in \partial W$. To handle this omission, it is desirable to extend the fixed point index to the class of "locally compact" mappings defined in metric ANR's. Such an extension was given independently by A. Granas [22] and F.E. Browder [8] in the early 1970's. Before describing this extension, we first indicate how, assuming one knows the properties of the classical fixed point index, one can define the Leray–Schauder degree. Conversely, we shall see later that, starting from the Leray–Schauder degree, one can obtain the fixed point index.

To define Leray–Schauder degree, we begin with some definitions. If X and Y are topological spaces, $V \subset X$ is an open subset of X and $f: V \longrightarrow Y$ is a continuous map, f is called "locally compact on V" if, for each $x \in V$, there exists an open neighborhood U_x of x such that $\overline{f(U_x)}$ is compact. If S is a compact subset of V and f is locally compact on V, it is easy to see that there exists an open neighborhood U of S, $U \subset V$, such that $\overline{f(U)}$ is compact.

Now suppose that X is a Banach space, V is an open subset of X and $f: V \longrightarrow X$ is a continuous map such that $S = \{x \in V \mid f(x) = x\}$ is compact (possibly empty). Assume that f is locally compact on some open neighborhood of S. Our previous remarks show tghat there is an open neighborhood U of S with $U \subset V$ and $\overline{f(U)}$ compact. If A is an compact subset of X and if we denote by $\overline{co}(A)$ the smallest, closed, convex set containing A, it is known that $\overline{co}(A)$ is compact. Thus there exists a compact metric ANR K with $\overline{f(U)} \subset K$, for example, $K = \overline{co}(f(U))$. We define $\deg_{LS}(I-f,V,0)$, the Leray–Schauder degree of $I-f$ on V with respect to 0, by

$$\deg_{LS}(I-f,V,0) = i_K(f,U \cap K).$$

If S is empty, U may be empty, and we understand $i_K(f,U \cap K) = 0$ then.

The obvious difficulty is to prove that this definition is independent of the particular open neighborhood U of S with $\overline{f(U)}$ compact and the particular compact metric ANR K with $\overline{f(U)} \subset K$. Thus suppose that U_j, $j=1,2$, is an open neighborhood of S with $\overline{f(U_j)}$ compact and that K_j is a compact metric ANR with $f(U_j) \subset K_j$. We must prove that

$$i_{K_1}(f, U_1 \cap K_1) = i_{K_2}(f, U_2 \cap K_2).$$

Let $U = U_1 \cap U_2$, $C_j = \overline{co}(K_j)$ and $C = C_1 \cap C_2$. Because all fixed points of f in $U_j \cap K_j$ lie in $U \cap K_j$, the additivity property of the fixed point index implies

$$i_{K_j}(f, U_j \cap K_j) = i_{K_j}(f, U \cap K_j).$$

Because $f(U \cap C_j) \subset K_j$, the commutativity property implies

$$i_{K_j}(f, U \cap K_j) = i_{C_j}(f, U \cap C_j).$$

Finally, C is a compact, convex set and hence a metric ANR and

$$f(U \cap C_j) \subset f(U_1) \cap f(U_2) \subset C_1 \cap C_2 = C,$$

so the commutativity property shows that

$$i_{C_j}(f, U \cap C_j) = i_C(f, U \cap C).$$

Thus we have proved that

$$i_{K_j}(f, U_j \cap K_j) = i_C(f, U \cap C), \; j=1,2,$$

which shows that $\deg_{LS}(I - f, V, 0)$ is well-defined.

Naturally, if f, V and X are as above, we also define $i_X(f,V)$ by

$$i_X(f,V) := \deg_{LS}(I - f, V, 0).$$

If X is a normed linear space, rather than a Banach space, we can still define $\deg_{LS}(I - f, V, 0)$, or, equivalently, $i_X(f,V)$. However, the previous argument must be modified, because it is no longer necessarily true that $\overline{co}(A)$ is compact when A is compact.

When X is a normed linear space $i_X(f,V)$ can be defined by finite dimensional approximation, which is the usual approach to Leray–Schauder degree [31]. We omit details and instead indicate how, in the Banach space case, the usual properties of Leray–Schauder degree follow directly from the corresponding properties of the classical fixed point index.

Theorem 1. (Additivity). Let V be an open subset of a normed linear space X and $f: V \to X$ a continuous map. If $\{x \in V \mid f(x) = x\}$ is empty, it follows that $i_X(f,V) := \deg(I-f,V,0) = 0$. If V_1 and V_2 are open subsets of V and $S_j = \{x \in V_j \mid f(x) = x\}$ is a compact subset of V_j for $j=1,2$, and f is locally compact on an open neighborhood of $S = S_1 \cup S_2$ and $S_1 \cap S_2$ is empty, it follows that

$$i_X(f,V_1 \cup V_2) = i_X(f,V_1) + i_X(f,V_2).$$

Proof. (When X is a Banach space). The fact that $i_X(f,V) = 0$ when f has no fixed points in V follows directly from the corresponding property for the classical fixed point index and from the definition of $i_X(f,V)$. Thus consider the case of V_1 and V_2. Let U be an open neighborhood of S with $\overline{f(U)}$ compact and define $U_j = U \cap V_j$, so U_j is an open neighborhood of S_j. If $K = \overline{\text{co}}f(U)$, the additivity property of the classical fixed point index implies

$$i_K(f,(U_1 \cup U_2) \cap K) = i_K(f,U_1 \cap K) + i_K(f,U_2 \cap K).$$

However, our definition yields

$$i_X(f,V_1 \cup V_2) = i_K(f,(U_1 \cup U_2) \cap K) \quad \text{and}$$

$$i_X(f,V_j) = i_K(f,U_j \cap K) \quad \text{for} \quad j=1,2,$$

which gives Theorem 1. ■

It is equally easy to derive the homotopy property for the Leray–Schauder degree from the corresponding result for the classical fixed point index.

Theorem 2. (Homotopy). Let X be a normed linear space and Ω an open subset of $X \times [0,1]$. Assume that $f: \Omega \to X$ is a continuous map and define $\Omega_t = \{x \mid (x,t) \in \Omega\}$ and $f_t(x) = f(x,t)$ for $x \in \Omega_t$, $0 \le t \le 1$. Let $S = \{(x,t) \in \Omega \mid f(x,t) = x\}$ and assume that S is compact and that f is locally compact on an open neighborhood of S in Ω. Then it follows that $i_X(f_t,\Omega_t)$ is constant for $0 \le t \le 1$.

Proof. (When X is a Banach space). Let W be an open neighborhood of S in $X \times [0,1]$ such that $\overline{f(W)}$ is compact and define $K = \overline{\text{co}}f(W)$. If $W_t = \{x \mid (x,t) \in W\}$ our definition implies that

$$i_X(f_t,\Omega_t) = i_K(f_t,W_t \cap K).$$

The homotopy property for the classical fixed point index implies that $i_K(f_t,W_t \cap K)$ is constant for $0 \le t \le 1$. ■

The commutativity property is not explicitly stated in most treatments of the Leray–Schauder degree; no mention of it can be found in [37], for example. Nevertheless, the commutativity property plays an important role in our generalizations of the fixed point index. Again, it follows easily from the corresponding result for the classical fixed point index.

Theorem 3. (Commutativity). Let V_j be an open subset of a normed linear space X_j, $j=1,2$. Suppose that $g_1\colon V_1 \to X_2$ and $g_2\colon V_2 \to X_1$ are continuous maps, so $g_2g_1\colon g_1^{-1}(V_2) \to X_1$ and $g_1g_2\colon g_2^{-1}(V_1) \to X_2$. Assume that $S_1 = \{x \in g_1^{-1}(V_2)\colon g_2(g_1(x)) = x\}$ is compact. Then it follows that $S_2 = \{y \in g_2^{-1}(V_1)\colon g_1(g_2(y)) = y\}$ is compact and $S_2 = g_1(S_1)$. If g_1 is locally compact on an open neighborhood of S_1, then

$$i_{X_1}(g_2g_1,g_1^{-1}(V_2)) = i_{X_2}(g_1g_2,g_2^{-1}(V_1)).$$

Proof. (When X_1 and X_2 are Banach spaces). We leave to the reader the proof that S_2 is compact and $S_2 = g_1(S_1)$. Let U_1 be an open neighborhood of S_1 such that $U_1 \subset g_1^{-1}(V_2)$ and $\overline{g_1(U_1)}$ is compact. Let U_2 be an open neighborhood of S_2 with $U_2 \subset g_2^{-1}(U_1)$. It is easy to see that $\overline{g_2g_1(U_1)}$ and $\overline{g_1g_2(U_2)}$ are compact. Define $K_2 = \overline{co}g_1(U_1)$ and $K_1 = \overline{co}g_2(K_2 \cap V_2)$. One can easily check that $g_1\colon U_1 \cap K_1 \to K_2$ and $g_2\colon V_2 \cap K_2 \to K_1$. By using the additivity and commutativity properties of the classical fixed point index, one sees that

$$i_{K_1}(g_2g_1,U_1) = i_{K_2}(g_1g_2,U_2).$$

It follows that

$$i_{X_1}(g_2g_1,g_1^{-1}(V_2)) = i_{X_2}(g_1g_2,g_2^{-1}(V_1)). \quad ■$$

It remains to give a version of the normalization property for the Leray–Schauder degree. For the moment we restrict ourselves to a simple version of the normalization property. More general versions will be given later.

Theorem 4. (Normalization). Let V be an open subset of a normed linear space X. Suppose that there exists a compact metric ANR A such that $f(V) \subset A \subset V$. Let f_A denote f considered as a map from A to A and let $L(f_A,A)$ denote the Lefschetz number of $f_A: A \to A$ (singular homology with rational coefficients). Then it follows that

$$i_X(f,V) = L(f_A,A).$$

Proof. By definition of $i_X(f,V)$ and the normalization property for the classical fixed point index we have

$$i_X(f,V) = i_A(f,V \cap A) = i_A(f,A) = L(f_A,A). \qquad \blacksquare$$

If f denotes the point map $f(x) = a$ for all $x \in V$, where $a \in V$, the compact metric ANR A can be taken to be a point $\{a\}$ and $L(f_A,A) = 1$. Thus in this case we find that

$$\deg_{LS}(I - f,V,0) = 1,$$

which is often called the normalization property for the Leray–Schauder degree.

With these preliminaries, we can extend the fixed point index to the class of locally compact mappings defined on open subsets of metric ANR's. The basic idea of the generalization is to use the commutativity property in conjunction with certain imbedding and retraction theorems. This trick can be traced to J. Leray [33], who used it for compact metric ANR's. The same method was also employed by Deleanu [12]. Later, the idea was exploited by F.E. Browder [8] and A. Granas [23].

We need the following imbedding theorem.

Theorem 5. (Arens and Eells [1]). Let X be a metric space. There exists a normed linear space E and an imbedding j of X into E such that $j(X)$ is a closed subset of E and j is a homeomorphism of X onto $j(X)$.

If X in Theorem 5 is a metric ANR and we write $A = j(X)$, a closed subset of a metric space E, and $f = j^{-1}: A \to X$, the definition of ANR implies that there is an open neighborhood O of $j(X)$ in E and a continuous map $\rho: O \to X$ such that $\rho(y) = j^{-1}(y)$ for all $y \in j(X)$. In particular we have $\rho(j(x)) = x$ for all $x \in X$.

Now suppose that X is a metric ANR, V is an open subset of X and $f: V \to X$ is a continuous map such that $S = \{x \in V | f(x) = x\}$ is compact and f is locally compact on some open neighborhood of S. Let j be an imbedding of X as a closed subset $j(X)$ of a normed linear space E. Let O be an open neighborhood of $j(X)$ in E and $\rho: O \to X$ a continuous map which is an extension of $j^{-1}: j(X) \to X$. We <u>define</u> $i_X(f,V)$, the fixed point index of $f: V \to X$, by

$$i_X(f,V) := i_E(jf\rho, \rho^{-1}(V)).$$

Obviously, for this definition to make sense we must prove that it is independent of the particular imbedding j and the particular extension ρ of j^{-1}. Furthermore, we must show that our definition agrees with the usual definition when X is a compact metric ANR.

We consider the second point first. If X is a compact metric ANR, $Y = j(X)$ is a compact metric ANR. Since $(jf\rho)(\rho^{-1}(V)) \subset Y$, the definition of $i_E(jf\rho, \rho^{-1}(V))$ implies that

$$i_E(jf\rho, \rho^{-1}(V)) = i_Y(jf\rho, \rho^{-1}(V) \cap Y) = i_Y(jfj^{-1}, j(V)).$$

If we define $g_2(y) = j^{-1}(y)$ for $y \in Y$, so $g_2: Y \to X$ and $g_1(x) = j(f(x))$ for $x \in V$, so $g_1: V \to Y$, the commutativity property implies that

$$i_Y(jfj^{-1}, j(V)) = i_Y(g_1 g_2, g_2^{-1}(V)) = i_X(g_2 g_1, V) = i_X(f,V).$$

Thus we see that, for X a compact metric ANR,

$$i_E(jf\rho, \rho^{-1}(V)) = i_X(f,V).$$

Next, suppose that X is a metric ANR and that j_k is a homeomorphism of X onto a closed subset $j_k(X)$ of a normed linear space E_k for $k=1,2$. Let O_k be an open neighborhood of $j_k(X)$ for which there is a continuous map $\rho_k: O_k \to X$ which is an extension of j_k^{-1}. Suppose that V is an open subset of X and $f: V \to X$ is a continuous map with $S = \{x \in V | f(x) = x\}$ compact and f locally compact on an open neighborhood of S. We must prove that

$$i_{E_1}(j_1 f \rho_1, \rho_1^{-1}(V)) = i_{E_2}(j_2 f \rho_2, \rho_2^{-1}(V)).$$

However, proving this is just an application of Theorem 3. Define $X_k = j_k(X)$, $k=1,2$, and define $g_1 = j_2 f \rho_1 \colon \rho_1^{-1}(V) \to X_2$ and $g_2 = j_1 \rho_2 \colon O_2 \to X_1$. Because $(\rho_2 j_2)(x) = x$ for all $x \in X$, we see that $g_2 g_1 \colon \rho_1^{-1}(V) \to X_1$ and $g_2 g_1 = j_1 f \rho_1$. Similarly, we see that $g_1 g_2 \colon \rho_2^{-1}(V) \to X_2$ and $g_1 g_2 = j_2 f \rho_2$. Thus Theorem 3 yields the desired result.

Once one has shown that the definition of $i_X(f,V)$ is meaningful, it is relatively easy to derive the analogues of Theorems 1–4 in this generality. For completeness, we state the additivity, homotopy, commutativity and normalization properties, but we leave the proofs to the reader.

Theorem 6. (Additivity). Let X be a metric ANR, V an open subset of X and $f\colon V \to X$ a continuous map. If f has no fixed points in V, $i_X(f,V) = 0$. If V_1 and V_2 are open subsets of V and $S_k = \{x \in V_k \mid f(x) = x\}$ is compact and f is locally compact on an open neighborhood of S_k for $k=1,2$ and $S_1 \cap S_2$ is empty, then

$$i_X(f, V_1 \cup V_2) = i_X(f, V_1) + i_X(f, V_2).$$

Theorem 7. (Homotopy). Let X be a metric ANR, Ω an open subset of $X \times [0,1]$ and $f\colon \Omega \to X$ a continuous map with $\Sigma = \{(x,t) \in \Omega \mid f(x,t) = x\}$ compact and f locally compact on a neighborhood of Σ. Then it follows that $i_X(f_t, \Omega_t)$ is constant for $0 \le t \le 1$, where $\Omega_t = \{x \mid (x,t) \in \Omega\}$ and $f_t(x) = f(x,t)$ for $x \in \Omega_t$.

Theorem 8. (Commutativity). Let X and Y be metric ANR's with open subsets U and V respectively. Assume that $f\colon U \to Y$ and $g\colon V \to X$ are continuous maps and that $S = \{x \in f^{-1}(V) \colon g(f(x)) = x\}$ is compact and f is locally compact on a neighborhood of S. Then $T = \{y \in g^{-1}(U) \colon f(g(y)) = y\}$ is compact and

$$i_X(gf, f^{-1}(V)) = i_Y(fg, g^{-1}(U)).$$

Theorem 9. (Normalization). Let X be a metric ANR, V an open subset of X and $f\colon V \to X$ a continuous map. Assume that there exists a compact metric ANR A with $f(V) \subset A \subset V$. If f_A denotes $f|A$, viewed as a map from A to A, and $L(f_A,A)$ denotes the Lefschetz number of f_A, then it follows that

$$i_X(f,V) = L(f_A,A).$$

2. Generalizing the fixed point index and asymptotic fixed point theory.

While the fixed index for locally compact maps defined in metric ANR's is adequate for many applications, there are also many operators which arise naturally in theory or in applications and which are not locally compact. Even in discussing linear operators, noncompact mappings arise naturally: see Section 2 of [52] and [53]. In view of such examples, it is natural to try to extend the fixed point index to include as large a class of mappings as possible. The motivation here is both theoretical and practical: extending the fixed point index provides a tool for proving fixed point theorems, many of which have proved useful in analysis. Indeed, generalizing the fixed point index can be viewed as generalizing the Lefschetz fixed point theorem.

We begin our discussion of generalizing the fixed point index by describing a beautiful fixed point theorem of G. Darbo [11]. If (X,d) is a metric space and A is a subset of finite diameter in X, Kuratowski [30] has introduced the idea of the "measure of noncompactness of A", $\gamma(A)$:

$$\gamma(A) = \inf\{r>0 \mid A = \bigcup_{i=1}^{m} A_i, \text{ where } m<\infty \text{ and } \operatorname{diameter}(A_i) \leq r\}.$$

Of course, if B is a subset of X, the diameter of B is defined by

$$\operatorname{diameter}(B) := \delta(B) = \sup\{d(x,y): x,y \in B\}.$$

It is easy to see that if $A \subset B$, $\gamma(A) \leq \gamma(B)$, and that $\gamma(A_1 \cup A_2) = \max(\gamma(A_1), \gamma(A_2))$. If (X,d) is a complete metric space, a point set topology result tells us that $A \subset X$ has compact closure if and only if A is totally bounded; and this implies that if (X,d) is complete and $A \subset X$, then $\gamma(A) = 0$ if and only if $\overline{A}$ is compact.

If X is a topological space and $A_k \subset X$, $k \geq 1$, is a decreasing sequence of nonempty compact sets, it is known that $A_\infty = \bigcap_{k\geq 1} A_k$ is compact and nonempty. Furthermore, if U is any open neighborhood of A_∞, there exists an integer $k(U)$ such that $A_k \subset U$ for $k \geq k(U)$. Kuratowki used his measure of noncompactness to generalize this theorem.

Theorem 1. (Kuratowski [30]). Let (X,d) be a complete metric space and let $\gamma(A)$ denote the measure of noncompactness of $A \subset X$. If $A_k \subset X$ is a decreasing sequence of closed, nonempty sets and if $\lim_{k\to\infty} \gamma(A_k) = 0$, then $A_\infty := \bigcap_{k\geq 1} A_k$ is a compact, nonempty set. If U is any open neighborhood of A_∞, there exists an integer m_U such that $A_k \subset U$ for all $k \geq m_U$.

G. Darbo observed [11] that if (X,d) is a normed linear space (so d is a norm) then γ has additional properties which make it very useful: If X is a normed linear space and A is a bounded subset of X, let (as in Section 1) $\overline{co}(A)$ denote the smallest closed convex set containing A. Darbo [11] observed that $\gamma(\overline{co}(A)) = \gamma(A)$. Also, if A and B are bounded subsets of X and $A + B := \{a+b: a\in A, b\in B\}$, then

$$\gamma(A+B) \leq \gamma(A) + \gamma(B).$$

We shall define a "generalized measure of noncompactness" β on a Banach space X to be a map which assigns to each bounded subset $A \subset X$ a nonnegative number $\beta(A)$ and which has the following properties:

1) $\beta(A) = 0$ if and only if $\overline{A}$ is compact.
2) For all bounded sets $A \subset X$, $\beta(A) < \infty$ and $\beta(\overline{co}(A)) = \beta(A)$.
3) For all bounded sets A and B in X, $\beta(A+B) \leq \beta(A) + \beta(B)$).
4) For all bounded sets A and B in X, $\beta(A \cup B) = \max(\beta(A), \beta(B))$.

If X is a Banach space and $L: X \to X$ is any bounded linear map and (for fixed $N<\infty$) we define

$$\beta(A) = \gamma(A) + \sum_{j=1}^{N} \gamma(L^j(A)),$$

one can prove that β is a generalized measure of noncompactness: see [52], p. 57. If (M,d) is a compact metric space and $X = C(M)$ is the space of continuous real-valued functions $u: M \to \mathbb{R}$ with $\|u\| = \sup_{t\in M} |u(t)|$, one can define a generalized measure of noncompactness which measures lack of equicontinuity. For $r>0$ and $A \subset X$, define $\beta_r(A) = \sup\{|u(s) - u(t)|: u \in A, d(s,t) \leq r\}$ and $\beta(A) = \lim_{r\to 0^+} \beta_r(A)$. One can prove that β is a generalized measure of noncompactness.

If (X_1,d_1) and (X_2,d_2) are metric spaces and $f: D_f \subset X_1 \to X_2$ is a continuous map, f is called a "k–set–contraction" if, for every set $A \subset D_1$ with $\gamma_1(A) < \infty$, $\gamma_2(f(A)) \leq k\gamma_1(A)$. (Here γ_1 and γ_2 are the Kuratowski measures of noncompactness in X_1 and X_2 respectively). If X_j is a subset of a Banach space Y_j and β_j is a generalized measure of noncompactness on Y_j $(j=1,2)$, then we say f is a "k–set–contraction (with respect to β_1 and β_2)" if $\beta_2(f(A)) \leq k\beta_1(A)$ for all $A \subset D_1$ with $\beta_1(A) < \infty$. If X_1 and X_2 are Banach spaces and $g: D_1 \subset X_1 \to X_2$ is a Lipschitz map with Lipschitz constant k and $h: D_1 \to X_2$ takes bounded sets to sets with compact closure, then $f = g+h$ is a k–set–contraction. Other examples are discussed in Section 1 of [40]. If $X_1 = X_2$ above, we shall usually assume that $\beta_1 = \beta_2$. If f is a k–set–contraction and $k<1$ we shall say that f is a "strict–set–contraction". If $f: D_1 \subset X_1 \to X_2$ and for each $x \in D_1$ there is an open neighborhood N_x of x and a constant $k_x < 1$ such that $f|N_x \cap D_1$ is a k_x–set–contraction, we shall say that f is a "local strict–set–contraction". From the viewpoint of generalizing the fixed point index, a natural assumption will be that f has a compact fixed point set S and that there is an open neighborhood U of S such that $f|U$ is a local strict–set–contraction. Further discussion of this point can be found in [40] and in [52], pp. 31–32. We remark that if M is a compact subset of a Banach space E, W is an open neighborhood of M and $f: W \to E$ is a C^1 map such that $f'(x)$ is a k–set–contraction for every $x \in M$, then for every $\varepsilon>0$ there is an open neighborhood V_ε of M for which $f|V_\varepsilon$ is a $(k+\varepsilon)$–set–contraction. Thus, if $k<1$, we are in the framework described above. Note, however, that if $k=0$ (so $f'(x)$ is compact for all $x \in M$) f may still fail to be compact on any open neighborhood of M: see [52], p. 33, for a simple example.

With these preliminaries we can state a slight variant of Darbo's theorem.

Theorem 2. (Darbo [11]). Let G be a closed, bounded, convex set in a Banach space X. Assume that $f: G \to G$ is a continuous map and that f is a k–set–contraction, $k>1$, with respect to a generalized measure of noncompactness β on G. Then f has a fixed point in G.

Proof. Fix $x_0 \in G$ and define $G_1 = \overline{co}(\{x_0\} \cup f(G))$ and $G_{n+1} = co(\{x_0\} \cup f(G_n))$ for $n \geq 1$. One can easily check that G_n, $n \geq 1$, is a decreasing sequence of closed, bounded convex sets with $x_0 \in G_n$, $f(G_n) \subset G_{n+1}$

and $\beta(G_n) \le k^n\beta(G)$ for all $n \ge 1$. If we define $G_\infty = \bigcap_{n\ge1} G_n$, it follows that $x_0 \in G_\infty$, G_∞ is compact and convex and $f(G_\infty) \subset G_\infty$. Thus the Schauder fixed point theorem implies that f has a fixed point in G_∞. ■

Sadovskii [59] has observed that Theorem 2 remains true if we assume only that $f: G \to G$ is continuous and that $\beta(f(A)) < \beta(A)$ whenever $A \subset G$ and $\beta(A) > 0$. See also Section 1 of [40] for a different proof of this fact.

Corollary 1. Let G be a closed, bounded convex subset of a Banach space X. Assume that $g: G \to X$ is a Lipschitz map with Lipschitz constant $k<1$, so $\|g(x) - g(y)\| \le k\|x-y\|$ for all $x,y \in G$. Assume that $h: G \to X$ is a compact map and that $f(x) = g(x) + h(x) \in G$ for all $x \in G$. Then it follows that f has a fixed point in G.

It is interesting to note that Corollary 1 has never been proved in the stated generality without using the ideas of Theorem 2.

It is natural, in view of Theorem 2, to try to extend the fixed point index to include k–set–contractions with $k<1$, and this has, in fact, been done in [38] and [40]. To describe this extension, we need some further definitions. Suppose that X is a closed subset of a Banach space E. We shall write $X \in \mathscr{F}$ and say that X is a locally finite union of closed, convex sets if there exists a collection C_j, $j \in J$, of closed, convex sets $C_j \subset E$ such that $X = \bigcup_{j\in J} C_j$ and $\{C_j: j \in J\}$ is a locally finite covering of X. We shall write $X \in \mathscr{F}_0$ and say that X is a finite union of closed, convex sets if there exist finitely many closed, convex sets $C_j \subset E$, $1 \le j \le m$, such that $X = \bigcup_{j=1}^{m} C_j$. If $X \in \mathscr{F}$, we shall view X as a metric space (in the metric inherited from E); a subset V of X is open if it is open as a subset of the topological space X.

Now suppose that $X \in \mathscr{F}$, V is an open subset of X and $f: V \to X$ is a continuous map. Assume that $S = \{x \in V \mid f(x) = x\}$ is compact and that there is an open neighborhood of U of S such that $f|U$ is a local strict–set–contraction. Then an integer $i_X(f,V)$, the fixed point index of $f: V \to X$, is defined in [40]. Direct generalizations of the additivity, homotopy, commutativity and normalization properties of the fixed point index for locally compact maps (see Section 1) are given in [38] and [40]. We omit further details here, since we shall later describe a still more general version of the fixed point index.

It is not hard to show that if $X \in \mathscr{F}$, then X is a metric ANR, and the metric on X is complete. However, it is not yet known whether a fixed point index can be defined for local strict–set–contractions defined on open subsets of general metric ANR's X (assuming the metric on X is complete).

There are other interesting classes of maps for which one can prove fixed point theorems but which are not necessarily local strict–set–contractions. The following theorem was originally claimed by R.L. Frum–Ketkov [20] for the case of balls in a Banach space. However, it is not hard to show that Frum–Ketkov's proof is incorrect. A proof for the special case of a ball in a so–called π_1–space was given in Theorem 2, p. 192 of [42]; a much more general version of the theorem can be found in Theorem 2, p. 370 of [43]. See also Corollaries 1 and 2 on pp. 353–354 of [43].

Theorem 3. Let G be a closed, bounded, convex subset of a Banach space. Let $f: G \to G$ be a continuous map. Assume that there exists a compact set $M \subset G$ and a constant $k<1$ such that

$$d(f(x),M) \leq kd(x,M)$$

for all $x \in G$. (Here $d(y,M) := \inf\{\|y-z\|: z \in M\}$). Then it follows that f has a fixed point in M.

We shall see later that the hypotheses in Theorem 3 can be greatly weakened. But for the moment we only consider Theorem 3 as providing another example of a class of maps for which a fixed point index should be defined.

Before stating our next result, we need a definition.

Definition 1. (See [43], p. 354). Let X be a topological space and $f: X \to X$ a continuous map. If M is a subset of X, we say that M is an "attractor for compact sets under f" if (1) M is compact, nonempty and $f(M) \subset M$ and (2) given any compact set $K \subset X$ and any open neighborhood U of M, there exists an integer $N = N(U,K)$ such that $f^n(K) \subset U$ for all $n \geq N$.

One can prove that, under the hypotheses of Theorem 2, there exists an attractor for compact sets under f. In fact, the set G_∞ in the proof of Theorem 2 is an attractor for compact sets under f.

With the above definition we can state our next result, which is Corollary 9 on p. 367 in [43]. More general versions of the following theorem can be found in [49]: see Proposition 2.4, Lemmas 3.1 and 3.2 and Proposition 3.3 in [49].

<u>Theorem 4</u>. (See Corollary 9, p. 367 in [43]). Let G be a closed, convex set in a Banach space E and $f: G \longrightarrow G$ a continuous map. Assume that the interior of G is nonempty and that there exists a set $M \subset \mathring{G}$ which is an attractor for compact sets under f. Finally, suppose that there exists an open neighborhood V of M and an integer $n \geq 1$ such that $f^n | V$ is a k–set–contraction, $k>1$, and $f|V$ is continuously Fréchet differentiable. Then f has a fixed point in M.

The key point to observe in Theorem 4 is that we do not know that $f|V$ is a k–set–contraction for some $k<1$. We only know that $f|V$ is C^1 and that $f^n|V$ is a k–set–contraction for some $n \geq 1$ and some k, $0 \leq k < 1$. If $n>1$, the fact that $f|V$ is C^1 plays a crucial role in the proof. Theorem 4 appears substantially harder that Theorem 2, and there are no known simple proofs (see [43], [49], [17] and [62]).

We now wish to describe how the fixed point index can be generalized to include maps like those in Theorems 2, 3 and 4. As a first step, we follow J. Leray [34] and generalize the Lefschetz number. If V is a vector space and $T: V \longrightarrow V$ is a linear endomorphism, let $N = \{x \in V: T^n(x) = 0 \text{ for some } n \geq 1\}$. Clearly, $T(N) \subset N$, so T induces a map $\bar{T}: W \longrightarrow W$, where $W = V/N$. If W is finite dimensional, Leray [34] defines the generalized trace of T, $tr_{gen}(T)$, by the formula

$$tr_{gen}(T) = tr(\bar{T}).$$

It is easy to show this definition agrees with the usual definition of trace when V is finite dimensional. A slightly different viewpoint (used in [9]) is sometimes useful. One can prove (see [40]) that V/N is finite dimensional if and only if there exists a finite dimensional subspace E of V with the properties that (1) $T(E) \subset E$ and (2) for every $x \in V$ there exists $m = m(x)$ with $T^m(x) \in E$. For any such subspace E, $tr_{gen}(T) = tr(T|E)$.

Next let X be a topological space, $f\colon X \to X$ be a continuous map and $H_i(X)$ denote the ith singular homology group over the rationals (so $H_i(X)$ is a vector space over the rationals). Let $f_{*i}\colon H_i(X) \to H_i(X)$ denote the usual homology map induced by f, so f_{*i} is a vector space endomorphism. If $tr_{gen}(f_{*i})$ is defined for all $i \geq 0$ and 0 except for finitely many i, Leray [34] defines the generalized Lefschetz number, $L_{gen}(f,X)$ or $L_{gen}(f)$ by the formula

$$L_{gen}(f) = \sum_{i \geq 0} (-1)^i tr_{gen}(f_{*i}).$$

Obviously, if $f,g\colon X \to X$ and f and g are homotopic, then $L_{gen}(g)$ is defined if and only if $L_{gen}(f)$ is defined and $L_{gen}(f) = L_{gen}(f)$. If $Y \subset X$ and $f(X) \subset Y$, one can prove that $L_{gen}(f)$ is defined if and only if $L_{gen}(f|Y)$ is defined and $L_{gen}(f) = L_{gen}(f|Y)$. More generally, if $Y \subset X$ and $f(Y) \subset Y$ and $f^n(X) \subset Y$ for some $n \geq 1$, then $L_{gen}(f)$ is defined if and only if $L_{gen}(f|Y)$ is defined and $L_{gen}(f) = L_{gen}(f|Y)$. Simple proofs of these facts are outlined in [43], pp. 351–352.

The above facts about the generalized Lefschetz number remain true no matter what homology theory is used. The following result of G. Fournier [19] uses the fact that singular homology is compactly supported.

Lemma 1. (See [19]). Let X be a topological space, $f\colon X \to X$ a continuous map and $Y \subset X$ a subset of X such that $f(Y) \subset Y$. Assume that for every compact set $K \subset X$, there exists an integer $n = n(K)$ such that $f^n(K) \subset Y$. Then the generalized Lefschetz number $L_{gen}(f)$ is defined if and only if $L_{gen}(f|Y)$ is defined, and $L_{gen}(f|Y) = L_{gen}(f)$. (Here $f|Y$ denotes the restriction of f to Y, considered as map of Y to Y).

The generalized Lefschetz number alone cannot give us fixed point theorems without some restrictions on the map f. For example, if V denotes the closed unit ball in an infinite dimensional Banach space E, it is known that for every

$\varepsilon>0$, there exists a continuous map $f: V \to V$ such that $\inf\{\|f(x)-x\|: x \in V\} \geq 1-\varepsilon$. (To see this, recall that there is a retraction r of V onto $\partial V = \{x \mid \|x\| = 1\}$, so ∂V is an AR. Define $f(x) = -x/\|x\|$ for $\|x\| \geq \varepsilon$. Because ∂V is an AR, f can be extended as a continuous map $f: V \to \partial V$, and one can check that $\|f(x) - x\| \geq 1-\varepsilon$ for all $x \in V$. Nevertheless, because f is homotopic to a point map, $L(f) = 1$.)

To prove fixed point theorems and develop a generalized fixed point theorem, we need some geometrical results about spaces $X \in \mathscr{F}_0$. The next lemma, which can be viewed as a generalization of Dugundji's theorem that a convex subset of a normed linear space is an AR, will play a crucial role in our analysis. A proof is given in [40].

Lemma 2. (See [40, 43]). Let C and D be closed subsets of a Banach space E. Assume that $C = \bigcup_{j=1}^{m} C_j$ and $D = \bigcup_{j=1}^{m} D_j$, where C_j and D_j are closed, convex subsets of E for $1 \leq j \leq m$ and $D_j \subset C_j$ for $1 \leq j \leq m$. For every subset J of $\{1,2,\cdots,m\}$, define $C_J = \bigcap_{j \in J} C_j$ and $D_J = \bigcap_{j \in J} D_j$ and assume that C_J is nonempty if and only if D_J is nonempty. Then there exists a continuous retraction R of C onto D such that $R(C_j) \subset C_j$ for $1 \leq j \leq m$.

Using Lemma 2, it is easy to prove the following variant, whose proof is implicit in the proof of Theorem 1 in [43] and which is stated explicitly as Lemma 1.1 in [49].

Lemma 3. (See [43] and [49]). Assume that C is a closed subset of a Banach space E and that $C = \bigcup_{j=1}^{m} C_j$, where C_j is a closed, convex subset of E for $1 \leq j \leq m$. If A is any compact subset of C, there exists a continuous map $R: C \to C$ such that $R(C)$ has compact closure, $R(C_j) \subset C_j$ for $1 \leq j \leq m$ and $R(a) = a$ for all $a \in A$. If $\gamma(C) < \delta$, where γ denotes Darbo's measure of noncompactness, then R can also be chosen so that $\|R(x) - x\| < \delta$ for all $x \in C$.

Suppose $C = \bigcup_{j=1}^{m} C_j$, where C_j is a closed, convex subset of a Banach space E. If $J \subset \{1,2,\cdots,m\}$ and C_J is nonempty, select $x_J \in C_J$ and define $D_j = \overline{co}\{x_J: C_J$ is nonempty and $j \in J\}$. Obviously D_j is compact and convex and, in fact, finite dimensional. Lemma 2 implies that there is a retraction R of C onto $D = \bigcup_{j=1}^{m} D_j$ such that $R(C_j) \subset C_j$ for $1 \le j \le m$. If we consider the homotopy $H(x,t) = (1-t)R(x) + tx$, $0 \le t \le 1$, the fact that $R(C_j) \subset C_j$ implies that $H(x,t) \in C$ for $0 \le t \le 1$, and $x \in C$. Thus D is a deformation retract of C. Since D is a compact metric ANR (in fact, a finite dimensional simplicial complex), $H_i(D)$ is finite dimensional for all i and equals 0 except for finitely many i. Since $H_i(C)$ is isomorphic to $H_i(D)$, $H_*(C)$ is also finite dimensional.

Lemma 3 is directly relevant to fixed point theory, as we illustrate in the following theorem.

Theorem 5. Suppose that $C \in \mathscr{F}_0$ and $f: C \to C$ is a continuous map. If $L(f) \neq 0$ (where $L(f)$ denotes the Lefschetz number of f) and $\gamma(C) < \delta$ (where γ is the measure of noncompactness), then there exists $x_0 \in C$ such that $\|f(x_0) - x_0\| < \delta$.

Proof. Let R be as in Lemma 3, so $R(C)$ has compact closure, $R(C_j) \subset C_j$ for $1 \le j \le m$ and $\|R(x) - x\| < \delta$ for all x. If we write $D_j = \overline{co}R(C_j) \subset C_j$ and $D = \bigcup_{j=1}^{m} D_j$, $R(C) \subset D$ and D is a compact metric ANR. We know that f and Rf are homotopic as maps of C to C by $(1-t)f(x) + tRf(x)$, $0 \le t \le 1$, so $L(f) = L(Rf)$. However, we have $(Rf)(C) \subset D$, so $L(Rf) = L(Rf|D) \neq 0$. The Lefschetz fixed point theorem implies that Rf has a fixed point $x_0 \in D$, so $Rf(x_0) = x_0$. It follows that

$$\|f(x_0) - x_0\| = \|f(x_0) - R(f(x_0))\| < \delta. \quad \blacksquare$$

We now want to exploit the idea of Theorem 5 further in order to define an "approximate fixed point index" (see [49], Section 1). Suppose that $X \in \mathscr{F}_0$, X is a closed subset of a Banach space E, U is a bounded open subset of X (open in the relative topology in X, not open as a subset of E) and $f: \overline{U} \to X$ is a continuous map. Assume that

$$\inf\{\|f(x) - x\|: x \in \overline{U} - U\} = \delta > 0.$$

Definition 2. Let f, U, X, E and δ be as above and suppose that $X = \bigcup_{j=1}^{m} C_j$, where C_j is a closed, convex subset of E for $1 \le j \le m$. A continuous map $g: \overline{U} \to X$ is an "admissible approximation to f with respect to $\{C_j: 1 \le j \le m\}$" if (1) $g(\overline{U})$ is compact, (2) $\|f(x) - g(x)\| < \delta$ for all $x \in \overline{U}$ and (3) if $f(x) \in C_j$ for some j, then $g(x) \in C_j$.

In the context of Theorem 5, with $U = C$, $g = Rf$ is an admissible approximation to f with respect to $\{C_j: 1 \le j \le m\}$. Here $\overline{U} - U$ is empty and any number δ is allowed.

Notice that if g is as in Definition 2, then the classical fixed point index $i_X(g,U)$ is defined; and if $i_X(g,U) \neq 0$, $g(x_0) = x_0$ for some $x_0 \in U$ and $\|f(x_0) - x_0\| < \delta$. This suggests that we should define $i_X(f,U)$, the approximate fixed point index of f on U, by

$$i_X(f,U) = i_X(g,U).$$

However, before we can make this definition we must consider several points. It may happen that $X = \bigcup_{k=1}^{n} D_k$, where D_k is a closed, convex subset of E for $1 \le k \le n$. If h is an admissible approximation to f with respect to $\{D_k: 1 \le k \le n\}$ and g is as in Definition 3, it is unclear a priori whether

$$i_X(g,U) = i_X(h,U).$$

Furthermore, it is unclear whether there exists an admissible approximation to f with respect to $\{C_j: 1 \le j \le m\}$. The following proposition shows that neither of these problems arises.

Theorem 6. (See Proposition 1.1 in [49]). Suppose that $X \in \mathscr{F}_0$, X is a closed subset of a Banach space E, U is an open subset of X (in the relative topology on X) and $f: \overline{U} \to X$ is a continuous map. Assume that

$$\inf\{\|f(x) - x\|: x \in \overline{U} - U\} = \delta > 0$$

and $\gamma(f(U)) < \delta$, where γ denotes the measure of noncompactness. Then there exists a covering $\{C_j: 1 \le j \le m\}$ of X by closed, convex sets and a continuous mapping $g: \overline{U} \to X$ which is an admissible approximation to f with respect to

$\{C_j: 1 \le j \le m\}$. If $\{D_k: 1 \le k \le n\}$ is a covering of X by closed, convex sets D_k and $h: \overline{U} \to X$ is an admissible approximation to f with respect to $\{D_k: 1 \le k \le n\}$, it follows that

$$i_X(g,U) = i_X(h,U).$$

By virtue of Theorem 6, we can now define the approximate fixed point index.

Definition 3. Suppose that $X \in \mathcal{F}_0$, X is a closed subset of a Banach E, U is a bounded open subset of the topological space X and $f: \overline{U} \to X$ is a continuous map. Assume that

$$\inf\{\|x - f(x)\|: x \in \overline{U} - U\} = \delta > 0$$

and $\gamma(f(U)) < \delta$, where γ denotes the measure of noncompactness. Then, if g is an admissible approximation of f with respect to $\{C_j | 1 \le j \le m\}$, where $X = \bigcup_{j=1}^{m} C_j$ and C_j is closed and convex for $1 \le j \le m$, $i_X(f,U)$, the approximate fixed point index of f, is defined by

$$i_X(f,U) := i_X(g,U).$$

Our construction shows that if $i_X(f,U) \neq 0$, then there exists $x_0 \in U$ with $\|f(x_0) - x_0\| < \delta$ (δ as in Definition 3). Furthermore, Theorem 6 and the proof used in Theorem 5 show that if $X \in \mathcal{F}_0$ and $U = X$ (as $\overline{U} - U$ is empty) and $f: X \to X$ is a continuous map with $\gamma(f(X)) < \delta$, then

$$i_X(f,X) = L(f)$$

and there exists $x_0 \in X$ with $\|f(x_0) - x_0\| < \delta$. The other properties of the fixed point index all carry over to the approximate fixed point index. We refer the reader to Section 1 of [49] for further details.

Our interest in the approximate fixed point index is primarily as a tool with which to define a fixed point index which will include the examples of Theorems 2,3 and 4. We now present such a definition. However, we request the reader's indulgence: the definition may at first seem technical and unnatural. The justification for our approach is that (as we shall show later) it provides a unified framework for developing a fixed point index for many classes of maps – in particular the maps in Theorems 2,3, and 4.

Suppose that U is an open subset of a topological space $X \in \mathscr{F}$ and $f: U \longrightarrow X$ is a continuous map such that $S = \{x \in U: f(x) = x\}$ is compact (possibly empty). Suppose that there exists a bounded open neighborhood W of S with $\overline{W} \subset U$ and a decreasing sequence of spaces $K_n \in \mathscr{F}_0$, $n \geq 0$, such that (1) $K_0 \supset W$, (2) $f(W \cap K_n) \subset K_{n+1}$ for all $n \geq 0$ and (3) $\lim_{n\to\infty} \gamma(K_n) = 0$.

<u>Definition 4</u>. If the above conditions are satisfied for some $W \supset S$ and some decreasing sequence $K_n \in \mathscr{F}_0$, $n \geq 0$, then we say that "f belongs to the fixed point index class," and we define $i_X(f,U)$ by

$$i_X(f,U) = \lim_{n\to\infty} i_{K_n}(f, W \cap K_n).$$

Here $i_{K_n}(f, W \cap K_n)$ denotes the approximate fixed point index and is defined for all sufficiently large n.

It is proved in Proposition 2.1 of [49] that $i_X(f,U)$ is independent of the particular $W \supset S$ and the particular sequence $K_n \in \mathscr{F}_0$. Direct generalizations of the additivity, homotopy, commutativity and normalization properties are given in Proposition 2.1–2.4 of [49]. For example, if $i_X(f,U) \neq 0$, f has a fixed point in U. For reasons of length, we must omit a detailed discussion of most of these points. However, it will be useful to state the normalization property for our generalized fixed point index.

<u>Theorem 7</u>. (Compare Proposition 2.4, p. 265 in [49]). Suppose that Ω is an open subset of a topological space $X \in \mathscr{F}$ and that $f: \Omega \longrightarrow \Omega$ is a continuous map. Assume that there exists a compact $M \subset \Omega$ which is an attractor for compact sets under f (see Definition 1). Assume that there exists an open neighborhood W of M and a decreasing sequence of sets $K_n \in \mathscr{F}_0$ for $n \geq 0$ such that (1) $W \subset K_0 \subset \Omega$, (2) $f(W \cap K_n) \subset K_{n+1}$ for all $n \geq 0$ and (3) $\lim_{n\to\infty} \gamma(K_n) = 0$. Then for every integer $p \geq 1$ f^p belongs to the fixed point index class (see Definition 4), the generalized Lefschetz number of $f^p: \Omega \longrightarrow \Omega$ is defined and $i_X(f^p,\Omega) = L_{gen}(f^p)$. In particular, if $L_{gen}(f^p) \neq 0$, f^p has a fixed point in Ω.

<u>Proof</u>. We present an outline of the proof, given in Proposition 2.4 of [49], since

we shall need not only the statement of Theorem 7 but also some results which are obtained in the course of the proof.

By using Lemma 5 in [43], one sees that there exists an open neighborhood V of M with $V \subset W$ and $f(V) \subset V$. The existence of V depends only on the existence of an attractor for compact sets under f. Next, by using Lemma 9 in [43] one proves that there exists an open neighborhood U of M with $U \subset V$ and $\overline{f(U)} \subset U$. The existence of U depends only on the existence of the attractor M and the fact that $\lim_{j\to\infty} \gamma(f^j(V)) = 0$.

Once one has proved the existence of U, one proves (see [49], p. 265–266) that there exists a decreasing sequence of spaces $A_n \in \mathscr{F}_0$, $n\geq 0$, with (1) $U \subset A_0$, (2) $f(U \cap A_n) \subset A_{n+1}$ for all $n\geq 0$, (3) $\lim_{n\to\infty} \gamma(A_n) = 0$ and (4) $A_n \subset U$ for all sufficiently large n. If $A_n \subset U$ for $n\geq m$, it follows that

$$f(A_n \cap U) = f(A_n) \subset A_{n+1} \subset A_n.$$

It follows from the definition of the generalized index that

$$i_X(f,\Omega) = \lim_{n\to\infty} i_{A_n}(f, U \cap A_n) = \lim_{n\to\infty} i_{A_n}(f,A_n).$$

If we write f_{A_n} to denote $f\colon A_n \to A_n$ for $n\geq m$, the normalization property gives

$$i_{A_n}(f,A_n) = L(f_{A_n}).$$

If K is any compact subset of Ω, $f^j(K) \subset U$ for all $j \geq N(K,U)$ (because M is an attractor), so $f^j(K) \subset A_n$ for all $j \geq N(K,U) + n$. Fournier's lemma (see Lemma 1) now implies that

$$L_{gen}(f) = L(f_{A_n}).$$

If $p\geq 1$ and $g = f^p$, note that all fixed points of g lie in $M \subset U$ and that $g(U \cap B_j) \subset B_{j+1}$, where $B_j := A_{jp}$ for $j\geq 0$. It follows from our definition that the generalized fixed point index of $g = f^p$ is defined and

$$i_X(f^p,\Omega) = \lim_{j\to\infty} i_{B_j}(f^p, U \cap B_j)$$

$$= \lim_{j\to\infty} i_{B_j}(f^p, B_j).$$

If g_{B_j} denotes the map $f^p: B_j \to B_j$, the same argument given for f shows that for j large

$$L_{gen}(f^p) = L(g_{B_j}) = i_{B_j}(f^p, B_j) = i_X(f^p, \Omega).$$

In particular, if $L_{gen}(f^p) \neq 0$, f^p has a fixed point in Ω. ■

Krasnoselskii and Zabreiko [29] and Steinlein [58] and [59] have given a very useful result, often called the mod p theorem, which relates the fixed point index of f^p (p a prime) to the fixed point index of f. A different proof of the mod p theorem, along lines suggested by A. Tromba, is given in Section 3 of [52]. The basic difficulty in proving the mod p theorem is to establish it for maps defined on open subsets of $\mathbb{R}^n$. We state here (without proof) a version of the mod p theorem which is valid for the generalized fixed point index. We note only that the version given here can be derived from the mod p theorem for maps defined on open subsets of compact metric ANR's by a geometric argument using Lemmas 2 and
3. A proof of Theorem 8 below will be given in another paper.

Theorem 8. Suppose that Ω is an open subset of a topological space $X \in \mathcal{F}$ and that $f: \Omega \to X$ is a continuous map. If p is a prime, let Ω_1 denote the domain of f^p. Assume that $\Sigma := \{x \in \Omega_1: f^p(x) = x\}$ is compact (possibly empty) and that $f(\Sigma) = \Sigma$. Suppose that there exists an open neighborhood W of Σ, $\Sigma \subset W \subset \Omega$ and a sequence of space $K_n \in \mathcal{F}_0$, $n \geq 0$, such that (1) $W \subset K_0$, (2) $f(W \cap K_n) \subset K_{n+1}$ for all $n \geq 0$ and (3) $\lim_{n\to\infty} \gamma(K_n) = 0$. Then the generalized fixed point indices $i_X(f^p, \Omega_1)$ and $i_X(f, \Omega_1)$ are defined and

$$i_X(f^p, \Omega_1) \equiv i_X(f, \Omega_1) \pmod{p}.$$

In the special case that $\Omega = X$ and X is a compact metric ANR, the mod p theorem simply states that

$$L(f^p) \equiv L(f) \pmod{p}.$$

Relatively simple proofs are available when X is a compact polyhedron: see [54] and [16], p. 136.

The case of general compact metric ANR's can be obtained without much difficulty from the compact polyhedral case or proved directly [54].

If we assume the mod p theorem for the case of the Lefschetz number and compact metric ANR's, we can derive a useful result. We prove the following theorem by a geometrical argument which can be sharpened to give a proof of Theorem 8. A simpler argument can be given by working at the homology level, but the simpler argument cannot be generalized to give Theorem 8.

Theorem 9. Suppose that $X \in \mathscr{F}_o$ and $f: X \to X$ is a continuous map. If p is a prime, it follows that

$$L(f^p) \equiv L(f) \pmod{p}.$$

Proof. X is a closed subset of a Banach space E and $X = \bigcup_{j=1}^{m} C_j$, where C_j is a closed, convex set for $1 \leq j \leq m$. Lemma 3 implies that there is a compact space $Y \in \mathscr{F}_o$, $Y \subset X$, and a retraction R of X onto Y such that $R(C_j) \subset C_j$ for $1 \leq j \leq m$. If follows that $H(t,x) = (1-t)x + tR(x) \in X$ for all $x \in X$ and for $0 \leq t \leq 1$, and Y is a deformation retract of X. If we define $g = Rf$ and write g_Y to denote g viewed as a map from Y to Y, then, because Y is a compact metric ANR, we know that

$$L(g_Y^p) \equiv L(g_Y) \pmod{p}.$$

We know that f and g are homotopic by the homotopy $(1-t)f + tg$, so $L(f) = L(g)$; and $L(g) = L(g_Y)$, because $g(X) \subset Y$. The same argument shows that f^p is homotopic to gf^{p-1} by the homotopy $[(1-t)f + tg]f^{p-1}$. Next, we see that gf^{p-1} is homotopic to g^2f^{p-2} by the homotopy $g[(1-t)f + tg]f^{p-2}$. Continuing in this way we find that g^jf^{p-j} is homotopic in X to $g^{j+1}f^{p-j-1}$ for $0 \leq j < p$. Combining these homotopies we see that f^p is homotopic to g^p in X, so

$$L(f^p) = L(g^p) = L((g_Y)^p).$$

We conclude that

$$L(f^p) \equiv L((g_Y)^p) \equiv L(g_Y) \equiv L(f) \pmod{p}. \quad \blacksquare$$

By using Theorem 9 we can obtain more information from Theorem 7.

Theorem 10. Let notation and hypotheses be as in Theorem 7. If p is a prime number and $m \geq 1$ an integer, then

$$L_{gen}(f^{mp}) \equiv L_{gen}(f^m) \pmod{p}.$$

If there exists an integer N and a sequence of prime numbers p_i, $i \geq 1$, such that (a) $\lim_{i\to\infty} p_i = \infty$, (b) $L_{gen}(f^{p_i}) \neq 0$ and (c) $|L_{gen}(f^{p_i})| \leq N$, then f has a fixed point and $L_{gen}(f) \neq 0$.

Proof. The proof of Theorem 7 shows that, for fixed m and p, there exists $A = A_{jmp} \in \mathcal{F}_0$ with $f^m(A) \subset A$ and (writing f^m_A for $f^m: A \longrightarrow A$)

$$L_{gen}(f^m) = L(f^m_A) \text{ and } L_{gen}(f^{mp}) = L((f^m_A)^p).$$

Theorem 9 implies that

$$L(f^m_A) \equiv L((f^m_A)^p), \pmod{p}$$

which yields the first part of Theorem 10.

If there exists a sequence of primes p_i as above, select $p_j > N$. If $L_{gen}(f) = 0$, we obtain

$$L_{gen}(f^{p_j}) \equiv 0 \pmod{p_j}.$$

However, because we assume that $L_{gen}(f^{p_j}) \neq 0$, this implies that

$$|L_{gen}(f^{p_j})| \geq p_j > N,$$

a contradiction. Thus we must have that $L_{gen}(f) \neq 0$, and Theorem 7 implies that f has a fixed point. $\blacksquare$

As we have remarked earlier, the interest of results like Theorems 7,8 and 10 resides in the fact that various concrete classes of maps satisfy the hypotheses of these theorems. We begin by noting that the generalized fixed point index is defined for some interesting classes of maps.

Theorem 11. (See Propositions 3.1, 3.2 and 3.3 in [49]). Suppose that $X \in \mathscr{F}$, U is an open subset of X and $f: U \to X$ is a continuous map such that $S = \{x \in U \mid f(x) = x\}$ is compact (possibly empty). Assume that at least one of the following hypotheses holds:

(A) There is an open neighborhood W of S such that $f|W$ is a k–set–contraction with $k<1$.

(B) There exists a compact set M with $S \subset M \subset X$, a constant k with $0 \leq k < 1$ and a positive number r such that $d(f(x),M) \leq kd(x,M)$ whenever $x \in U$ and $d(x,M) < r$. (Here, $d(y,M)$ denotes the distance of y to M).

(C) X is a Banach space; and there is an open neighborhood W of S and an integer $N \geq 1$ such that $f|W$ is continuously Fréchet differentiable and $f^N|W$ is a k–set–contraction with $k<1$.

Then f belongs to the fixed point index class.

Considerations of length preclude a proof of Theorem 11, and we refer the reader to Section 3 of [49] for details. We remark, however, that proving Theorem 11 when hypothesis A or B holds is not difficult. The case of hypothesis C seems much harder, and the method of proof in this case goes back to Lemma 14 and Corollary 9 in [43]. See, also, Lemmas 3.1 and 3.2 and Proposition 3.3 in [49].

It still remains to give concrete conditions under which the hypotheses of Theorem 7 are satisfied. The proof of the following theorem follows from arguments in Section 3 of [49]; see, also, Lemma 14 and Corollaries 9 and 10 in [43].

Theorem 12. Suppose that Ω is an open subset of a topological space $X \in \mathscr{F}$ and that $f: \Omega \to \Omega$ is a continuous map. Assume that there exists a compact set M which is an attractor for compact sets under f (see Definition 1). Assume that at least one of the following hypotheses holds:

(A) There is an open neighborhood W of M such that $f|W$ is a k–set–contraction with $k<1$.

(B) There exists a compact set M_1, with $M \subset M_1 \subset \Omega$, a constant k with $0 \leq k < 1$ and a positive number r such that $d(f(x),M_1) \leq kd(x,M_1)$ whenever $x \in \Omega$ and $d(x,M_1) < r$.

(C) X is a Banach space and there is an open neighborhood W of M

and an integer $N \geq 1$ such that $f|W$ is continuously Fréchet differentiable and $f^N|W$ is a k–set–contradiction with $k<1$.

Then there exists an open neighborhood V of M and a decreasing sequence of sets $K_n \in \mathscr{F}_0$ for $n \geq 0$ such that (1) $V \subset K_0 \subset \Omega$, (2) $f(V \cap K_n) \subset K_{n+1}$ for all $n \geq 0$ and (3) $\lim_{n \to \infty} \gamma(K_n) = 0$ (where γ denotes measure of noncompactness). For every integer $m \geq 1$, f^m belongs to the fixed point index class, the generalized Lefschetz number of $f^m \colon \Omega \to \Omega$ is defined and $i_X(f^m, \Omega) = L_{gen}(f^m)$. If $L_{gen}(f^m) \neq 0$, f^m has a fixed point in Ω. If p is a prime,

$$L_{gen}(f^{mp}) \equiv L_{gen}(f^m) \pmod{p}.$$

If there exists an integer N and a sequence of prime numbers p_i, $i \geq 1$, with (a) $\lim_{i \to \infty} p_i = \infty$, (b) $L_{gen}(f^{p_i}) \neq 0$ and (c) $|L_{gen}(f^{p_i})| \leq N$, then $L_{gen}(f) \neq 0$ and f has a fixed point.

<u>Proof</u>. Theorem 12 follows directly from Theorems 7 and 10 if we can prove the existence of V and $K_n \in \mathscr{F}_0$, $n \geq 0$. As already remarked, the existence of V and K_n follows from arguments in Section 3 of [49]. The initial argument in the proof of Theorem 7 shows that if hypothesis A or C (respectively, hypothesis B) holds and Ω_1 is any open neighborhood of M (respectively, of M_1), then there is an open neighborhood Ω_2 of M (respectively, of M_1) with $\overline{f(\Omega_2)} \subset \Omega_2$. Using this observation, one can assume that the set W in hypothesis A or C also satisfies $\overline{f(W)} \subset W$ and $W \subset \Omega_1$. If one now argues exactly as in Propositions 3.1, 3.2 or 3.3 of [49] (depending on whether f satisfies hypothesis A, B or C, respectively), one obtains V and the sets $K_n \in \mathscr{F}_0$. ∎

It is easy to give reasonable conditions under which $\Lambda_{gen}(f) \neq 0$ and f has a fixed point. In the statement of the following result, recall that a topological space Y is said to be "contractible in itself to a point" if there exists a point $y_0 \in Y$ and a continuous homotopy $H \colon Y \times [0,1] \to Y$ such that $H(y,0) = y$ for all $y \in Y$ and $H(y,1) = y_0$ for all $y \in Y$. The homology of such a space Y is trivial ($H_i(Y) = 0$ for $i>0$ and $H_0(Y) = Q$), so any continuous map $g \colon Y \to Y$ satisfies $L(g) = 1$.

Corollary 2. Let hypotheses and notation be as in Theorem 12. Assume in addition the there exists a set $Y \subset \Omega$ and an integer $m_0 \geq 1$ such that (a) Y is contractible in itself to a point, (b) $f^j(Y) \subset Y$ for all $j \geq m_0$ and (c) for every compact set $K \subset \Omega$ there exists an integer $n(K)$ with $f^j(K) \subset Y$ for all $j \geq n(K)$. Then it follows that $L_{gen}(f^j) = 1$ for all $j \geq 1$ and f has a fixed point in Ω.

Proof. Fix $j \geq m_0$, write $g = f^j$ and denote by g_Y the map $g: Y \to Y$. Lemma 1 implies that

$$L_{gen}(g) = L_{gen}(f^j) = L_{gen}(g_Y) = 1 \quad \text{for} \quad j \geq m_0.$$

If $j \geq 1$, let p be a prime such that $pj \geq m_0$ and $p > |L_{gen}(f^j)| + 1$. Theorem 12 implies that

$$L_{gen}(f^j) \equiv L_{gen}(f^{pj}) = 1 \pmod{p},$$

and our choice of p then implies that

$$L_{gen}(f^j) = 1.$$

Since $L_{gen}(f) = 1$, Theorem 12 implies that f has a fixed point. ■

Corollary 2 immediately implies Theorem 4. To see this, let r be a continuous retraction of the Banach space E in Theorem 4 onto G and let $h = f \circ r$. Then $h: E \to G \subset E$ and all the hypotheses of Corollary 2 are satisfied by $h: E \to E$. (Take $\Omega = E$ and $Y = G$; Y is contractible in itself to a point because G is convex. If M is as in Theorem 4, M is still an attractor for compact sets under h, h is C^1 on an open neighborhood V of M and $h^N|V$ is a k-set-contraction, $k<1$).

The assumption in Theorem 12 that there exists a compact set M which is an attractor for compact sets under f is a natural assumption, especially in discussing the asymptotic behaviour of dissipative systems. See J. Hale's book [24]. We indicate one simple set of assumptions which imply that a map f possesses an attractor.

Proposition 1. Let X be a closed subset of a Banach space E and

$f: X \to X$ a continuous map. Assume that there exists a bounded set B and an integer N such that $f(B) \subset B$ and $f^N|B$ is a k-set-contraction with $k<1$. Assume also that for every compact set $K \subset X$ there exists an integer $n = n(K)$ with $f^n(K) \subset B$. Then there exists a compact set M which is an attractor for compact sets under f.

Proof. Define $B_j = \overline{f^j(B)}$ and note that B_j is a decreasing sequence of closed, nonempty sets and $\gamma(B_{jN}) \le k^j \gamma(B)$. Theorem 1 implies that $B_\infty := \cap B_j$ is compact and nonempty, and one can easily check that $f(B_\infty) \subset B_\infty$. Furthermore, if U is any open neighborhood of B_∞, Theorem 1 implies that $B_j \subset U$ for all j sufficiently large. If K is a compact subset of X, there exists an integer $n = n(k)$ with $f^n(K) \subset B$ and then $f^{n+j}(K) \subset f^j(B) \subset U$ for all j sufficiently large. Thus $M := B_\infty$ is an attractor for compact sets under f. ■

Note that the hypotheses of Proposition 1 will be satified if there exists an integer m such that $f^m(X) := B$ is bounded and there exists an integer N such that $f^N|B$ is a k-set-contraction with $k<1$. In particular, it suffices that there exist an integer m such that $\overline{f^m(X)}$ is compact.

Remark 1. Theorem 12 contains, as a very special case, Theorem 2 in [62], where it is assumed that Ω is an open subset of a Banach space E, that $T: \Omega \to \Omega$ is a C^1 map, and that $\overline{T^n(\Omega)}$ is a compact subset of Ω for some n. Furthermore, as we have tried to show, Theorem 12 is implicit in [49], which predates [62]. The real interest of Tromba's work in [61] and [62] is that it introduces some new approaches to earlier results.

Remark 2. Theorem 12 is motivated in a general sense by applications to analysis and by the problem of unifying and generalizing the Lefschetz fixed point theorem, Schauder's fixed point theorem, Darbo's theorem and other results. However, the immediate motivation comes from the following old conjecture, which was told to this author more than twenty-five years by Felix E. Browder.

Conjecture 1. Let G be a closed, bounded, convex set in a Banach space E and $f\colon G \to G$ a continuous map. Assume that there is an integer N such that $\overline{f^N(G)}$ is compact. Does f have a fixed point?

The answer to this question is not even known if $N=2$. If we define $M = \bigcap_{j \geq 1} \overline{f^j(G)}$, then M is an attractor for compact sets under f. If there exists an open neighborhood W of M in G (open in the relative topology on G) such that $f|W$ is a k-set-contraction with $k<1$ or if (the case $k=0$) $\overline{f(W)}$ is compact, then Corollary 2 implies that f has a fixed point. Similarly, if G has nonempty interior, $M \subset \overset{\circ}{G}$ and f is C^1 on an open neighborhood of M, then Corollary 2 again implies that f has a fixed point. One can give other conditions which imply that f has a fixed point, but all known results involve having some additional information about the behaviour of f on a neighborhood of M in G.

On the other hand, one might hope to find a counterexample to Conjecture 1. However, results like Corollary 2 suggest that any potential counterexample must have very unpleasant properties.

3. Periodic solutions of a differential-delay equation.

As an example of the use of the fixed point index, we shall study in this section the question of existence of periodic solutions of the differential-delay equation

$$x'(t) = -\lambda g(x(t-1)), \quad \lambda>0. \qquad (1)_\lambda$$

If g is C^1 near 0 and $ug(u) > 0$ for all $u \neq 0$, much progress has been made on this problem: see, for example, [44], [45], [46], [48] and the references there. However, we wish to study the less well-understood case when g is not differentiable at 0, a model example being

$$g(x) = \begin{cases} ax, & \text{for } 0 \leq x \\ bx, & \text{for } x<0 \,. \end{cases}$$

Before studying equation $(1)_\lambda$ we need to recall some further abstract results concerning "attractive fixed points" and "ejective fixed points". There is some confusion of terminology here, because what are called "attractive fixed points" in the asymptotic fixed point theory literature (see, for example, [46]) are called "uniformly asymptotically stable fixed points" on page 10 of Hale's book [24]. To bridge the gap, we use terminology from [24].

Suppose that X is a topological space, W is an open subset of X and $f: W \to X$ is a continuous map with a fixed point $x_0 \in W$. The fixed point $x_0 \in W$ is called a "stable fixed point of f" if, for every open neighborhood V of x_0, there is an open neighborhood U of x_0 such that $f^j(U) \subset V \cap W$ for all $j \geq 0$. The fixed point x_0 is said to "attract all points in a set W_1" if for every open neighborhood U of x_0 and every $x \in W_1$ there exists an integer $n = n(x,U)$ with $f^j(x) \in U$ for all $j \geq n(x,U)$. The fixed point x_0 is called "asymptotically stable" if (1) x_0 is stable and (2) for some open neighborhood W_1 of x_0, x_0 attracts all points in W_1. The fixed point x_0 is said to "attract a set $A \subset W$" if, for every open neighborhood U of x_0, there exists an integer $n(A,U)$ with $f^j(A) \subset U$ for all $j \geq n(A,U)$. The fixed point x_0 is called "uniformly asymptotically stable" if (1) x_0 is stable and (2) x_0 attracts some open neighborhood W_1 of x_0.

We now state a few simple point set topology lemmas.

Lemma 1. Let X be a topological space, W an open subset of X and $f: W \to X$ a continuous map which has a fixed point $x_0 \in W$. Assume that x_0 is an asymptotically stable fixed point of f. Then given any open neighborhood V of x_0, there exists an open neighborhood $U \subset V$ of x_0 such that $f(U) \subset U$. If K is a compact subset of W and $\bigcup_{j \geq 1} f^j(K) \subset W$, then the fixed point x_0 attracts K.

Proof. Let V be a given open neighborhood of x_0, $V \subset W$. By the definition of asymptotic stability, there exists an open neighborhood W_1 of x_0, $W_1 \subset W$, such that x_0 attracts all points in W_1. Because x_0 is a stable fixed point of f, there exists an open neighborhood V_1 of x_0, $V_1 \subset V \cap W_1$, such that $f^j(V_1) \subset V \cap W_1$, for all $j \geq 0$. Similarly, there exists an open neighborhood V_2 of x_0 such that $V_2 \subset V_1$ and $f^j(V_2) \subset V_1$ for all $j \geq 1$.

Define $U = \{x \in V_1: f^j(x) \in V_1$ for all $j \geq 1\}$. It is clear that $V_2 \subset U \subset V_1$ and $f(U) \subset U$. We have to prove that U is open. If $x \in U$, the definition of asymptotic stability implies that for each $x \in U$, there exists an integer $n(x)$ with $f^j(x) \in V_2$ for all $j \geq n(x)$. The continuity of f implies that there is an open neighborhood O_x of x with $f^j(y) \in V_1$ for $0 \leq j < n(x)$ and $f^{n(x)}(y) \in V_2$ for all $y \in O_x$. Since $f^j(V_2) \subset V_1$ for all $j \geq 1$, this implies that $f^j(y) \in V_1$ for all $y \in O_x$ and all $j \geq 1$, so that $O_x \subset U$. We have thus proved that U is open.

Next suppose that K is a compact subset of W and $\bigcup_{j \geq 1} f^j(K) \subset W$. If V_3 is an open neighborhood of x_0, we must prove that there is an integer N with $f^j(K) \subset V_3$ for all $j \geq N$. By the first part of the lemma, there exists an open neighborhood $U_3 \subset V_3$ of x_0 with $f(U_3) \subset U_3$. For each $x \in K_1$ there is an integer $n(x)$ with $f^j(x) \in U_3$ for all $j \geq n(x)$. By continuity of f and the invariance of U_3 under f, there exists an open neighborhood O_x of x with $f^j(y) \in U_3$ for all $j \geq n(x)$ and all $y \in O_x$. By compactness of K, we can cover K by a finite number of open sets O_{x_i}, $1 \leq i \leq m$, and then $f^N(K) \subset U_3$ for $N = \max\{n(x_i): 1 \leq i \leq m\}$. ■

If we slightly strengthen the hypotheses of Lemma 1, we can prove that the fixed point x_0 is uniformly asymptotically stable.

<u>Lemma 2</u>. Let notation and hypotheses be as in the statement of Lemma 1. Assume, in addition, that X is a complete metric space and that there exists an open neighborhood W_1 of x_0 such that $f^j(W_1)$ is defined for all $j \geq 1$ and $\lim_{j \to \infty} \gamma(f^j(W_1)) = 0$, where γ denotes Kuratowski's measure of noncompactness. Then the fixed point x_0 is uniformly asymptotically stable.

<u>Proof</u>. By Lemma 1, there exists a bounded open neighborhood U of x_0 with $\overline{U} \subset W_1$ and $f(U) \subset U$. Define $B_j = \overline{f^j(U)}$ and note that B_j is a decreasing sequence of closed, nonempty subsets of $\overline{U}$ and that $\gamma(B_j) \leq \gamma(f^j(W_1))$, so

$\lim_{j\to\infty} \gamma(B_j) = 0$. It follows from Theorem 1 in Section 2 that $M = \bigcap_{j\geq 1} B_j \subset \overline{U}$ is compact and nonempty. Since $\overline{f(B_j)} = B_{j+1}$, one can prove, with the aid of Theorem 1 in Section 2, that $f(M) = M$. However, Lemma 1 implies that x_0 attracts M, so it must be true that $M = x_0$. Theorem 1 in Section 2 then implies that if V is any open neighborhood of x_0, $B_j \subset V$ for all sufficiently large j, i.e., x_0 attracts U. ■

Our next result generalizes old theorems about the fixed point index of attractive fixed points: see [46] and Theorem 3.5 in Section 3 of [52].

Theorem 1. Suppose that $X \in \mathcal{F}$, W is an open subset of X, $f: W \to X$ is a continuous map and $x_0 \in W$ is an asymptotically stable fixed point of f and the only fixed point of f in W. Assume that f belongs to the fixed point index class. Then it follows that $i_X(f,W) = 1$.

Proof. Because f belongs to the fixed point index class, there exists an open neighborhood V of x_0 and a decreasing sequence of sets $K_n \in \mathcal{F}_0$, $n\geq 0$, $K_n \subset X$, such that (1) $V \subset K_0$, (2) $f(V \cap K_n) \subset K_{n+1}$ for all $n\geq 0$ and (3) $\lim_{n\to\infty} \gamma(K_n) = 0$. By Lemma 1, there exists an open neighborhood U of x_0, $\overline{U} \subset V$, with $f(U) \subset U$. It follows that $f(U \cap K_n) \subset U \cap K_{n+1}$ for all n, so $f^j(U) \subset U \cap K_j$ and $\lim_{j\to\infty} \gamma(f^j(U)) = 0$. Lemma 2 now implies that x_0 attracts U.

By definition, X is a closed subset of some Banach space E from which it inherits its metric, so, for $\varepsilon>0$, we can define $B_\varepsilon = \{x \in X: \|x-x_0\| \leq \varepsilon\}$. For any $\varepsilon>0$, we know that $f^j(U) \subset B_\varepsilon$ for all $j \geq j(\varepsilon)$, so

$$L_{gen}(f^j: U \to U) = L_{gen}(f^j: B_\varepsilon \to B_\varepsilon) \quad \text{for } j\geq j(\varepsilon).$$

If we can prove that B_ε is contractible in itself to a point for $0<\varepsilon\leq\varepsilon_0$, it follows from Theorem 7 in Section 2 that if $j \geq j(\varepsilon)$ and $0<\varepsilon\leq\varepsilon_0$,

$$i_X(f^j,W) = i_X(f^j,U) = L_{gen}(f^j: U \to U) = L_{gen}(f^j: B_\varepsilon \to B_\varepsilon) = 1.$$

It follows from Theorem 10 in Section 2 that for all sufficiently large primes p,

$$i_X(f,W) = L_{gen}(f: U \to U) \equiv L_{gen}(f^p: U \to U) = 1 \pmod{p},$$

and this implies that $i_X(f,W) = 1$.

Thus to complete the proof it suffices to prove that B_ε is contractible in itself to a point for small ε. Because $X \in \mathcal{F}$, there exists a locally finite covering of X by closed, convex sets $C_j \subset E$, $j \in J$. Using the local finiteness of the covering we see that there exists $\varepsilon_0 > 0$ such that $d(x_0, C_j) \geq \varepsilon_0$ for all $j \in J$ with $x_0 \notin C_j$. If $0 < \varepsilon < \varepsilon_0$, it follows that $B_\varepsilon \cap C_j$ is nonempty if and only if $x_0 \in C_j$. Using this fact, we find that if $0 < \varepsilon < \varepsilon_0$ and if we define $H(x,t) = (1-t)x + tx_0$ for $(x,t) \in B_\varepsilon \times [0,1]$, then $H: B_\varepsilon \times [0,1] \to B_\varepsilon$ and B_ε is contractible in itself to x_0. ■

We shall actually only use Theorem 1 in the special case that there is an open neighborhood W_1 of x_0 with $\overline{f(W_1)}$ compact, but we state a more general corollary of Theorem 1.

Corollary 1. Suppose that $X \in \mathcal{F}$, W is an open subset of X, $f: W \to X$ is a continuous map and $x_0 \in W$ is an asymptotically stable fixed point of f. Assume that one of the following conditions holds:

(A) There exists an open neighborhood $W_1 \subset W$ of x_0 such that $f|W_1$ is a k-set-contraction with $k<1$.

(B) X is a Banach space. There exists an open neighborhood $W_1 \subset W$ of x_0 and an integer $N \geq 1$ such that $f|W_1$ is continuously Fréchet differentiable and $f^N|W_1$ is a k-set-contraction with $k<1$.

Then if $W_2 \subset W$ is any open neighborhood of x_0 such that x_0 is the only fixed point of f in W_2, it follows that $i_X(f,W_2)$, the generalized fixed point index, is defined and $i_X(f,W_2) = 1$.

Proof. We can replace W by W_2 initially. Theorem 11 in Section 2 implies that $f|W_2$ belongs to the fixed point index class, so Corollary 1 follows immediately from Theorem 1. ■

Remark 1. It should be emphasized that Theorem 1 is a topological result and does not follow simply by linearizing at x_0. Even in case B of Corollary 1 it may happen that $f'(x_0)$ has eigenvalues λ with $|\lambda| = 1$.

Remark 2. Suppose that W is an open subset of a Banach space X and $f: W \to X$ is a continuous map which has a compact fixed point set

$S = \{x \in W | f(x) = x\}$. Assume that f is C^1 on some open neighborhood of S. Then there exists an open neighborhood U of S and an integer N such that $f^N|U$ is a k-set-contraction with $k<1$ if and only if the "essential spectral radius" of $f'(x)$ is less than one for each $x \in S$. (See [39] for a discussion of the essential spectral radius of a bounded linear operator L and for a discussion of the relation of this idea to facts about the spectrum of L). Alternately, there exists an open neighborhood U of S and integer N with $f^N|U$ a k-set-contraction, $k<1$, if and only if, for each $x \in S$, there exists an integer $n = n(x)$ such that $(f'(x))^n$ is a c-set-contraction, where $c = c(x)<1$. We omit the proofs of these facts.

F.E. Browder [56] has introduced the useful idea of an "ejective fixed point", a concept roughly opposite to that of an attractive fixed point. Suppose that X is a topological space, W is an open subset of X, $x_0 \in W$ and $f\colon W - \{x_0\} \to X$ is a continuous map. Then x_0 is called an "ejective point of f" (see [44], p. 267) if there exists an open neighborhood U of x_0 such that for each $x \in U - \{x_0\}$ there is an integer $m = m(x)$ such that $f^m(x)$ is defined and $f^m(x) \notin U$. A neighborhood U with these properties is an "ejective neighborhood of x_0". If f is defined and continuous at x_0 and $f(x_0) = x_0$, x_0 is called an "ejective fixed point of f". It may happen in applications that a map f has an ejective point x_0, but that f cannot be defined continuously at x_0: see Section 3 of [44].

A closed, convex subset G of a Banach space E is called "infinite dimensional" if there does not exist a finite dimensional linear subspace F of E with $G \subset F$. If there exists a finite dimensional linear subspace F with $G \subset F \subset E$, G is called finite dimensional. If G is a compact, convex infinite dimensional subset of a Banach space and $f\colon G \to G$ is a continuous map, F.E. Browder [6] has proved that f has a fixed point x_0 which is not ejective. If G is a compact, convex finite dimensional subset of a Banach space and $f\colon G \to G$ is a continuous map, then f has a fixed point x_0 such that either (a) x_0 is not ejective or (b) x_0 lies in the interior of G (where the interior of G is taken relative to the affine linear subspace spanned by G).

In [44], Section 1, it was observed that the local fixed point index at an ejective fixed point is frequently zero and that this fact implies the previously mentioned results concerning existence of nonejective fixed points. See Theorem 1.1, Corollary 1.1 and Corollary 1.2 in [44], pp. 268–270. For our further work we need information about the fixed point index at an ejective fixed point.

Theorem 2. (See Corollary 1.1 in [44]). Let G be a closed, convex and infinite dimensional subset of a Banach space E, W a relatively open subset of G, and $f: W \to G$ a continuous map which is a k–set–contraction, $k<1$, with respect to a generalized measure of noncompactness β. Assume that if $\beta(A_k) \to 0$ for any decreasing sequence of sets $A_k \subset G$, then $\gamma(A_k) \to 0$, where γ is Kuratowski's measure of noncompactness. Assume that $x_0 \in W$ is an ejective fixed point of f. Then if $U \subset W$ is any open neighborhood of x_0 (open in the relative topology on G) and x_0 is the only fixed point of f in U, it follows that $i_G(f,U) = 0$.

Proof. The additivity property of the fixed point index shows that $i_G(f,U)$ is independent of the particular neighborhood U in the statement of the theorem. For $\rho>0$ define $B_\rho = \{x \in G \mid \|x-x_0\| \le \rho\}$ and select ρ so small that $B_\rho \subset W$. By choosing U sufficiently small, we can assume that $U \subset B_\rho$ and $f(U) \subset B_\rho$ and $f(x) \neq x$ for all $x \in U - \{x_0\}$.

Define a retraction R of G onto B_ρ by

$$R(x) = \begin{cases} x, & \text{for } \|x-x_0\| \le \rho \\ (1-t_x)x_0 + t_x x, \; t_x := \rho\|x-x_0\|^{-1}, & \text{for } \|x-x_0\| > \rho . \end{cases}$$

It is easy to see that for any set $A \subset G$, $R(A) \subset \overline{\mathrm{co}}(\{x_0\} \cup A)$; and this implies that $\beta(R(A)) \le \beta(A)$. It follows that $g=Rf$ is a k–set–contraction, $k<1$, with respect to β and $g(G) \subset B_\rho$. By our construction we have that $g(x) = f(x)$ for all $x \in U$, so

$$i_G(f,U) = i_G(g,U).$$

Our construction also insures that $g(U) \subset B_\rho$, so the commutativity property of the fixed point index gives

$$i_G(g,U) = i_{B_\rho}(g,U).$$

Clearly B_ρ is closed, convex and bounded; and we leave to the reader the verification that B_ρ is infinite dimensional. Because g is defined on all of B_ρ

and is a k–set–contraction with $k<1$, we can consider g a map of B_ρ to B_ρ and apply Corollary 1.1 of [44] to obtain

$$i_{B_\rho}(g,U) = 0.$$

This completes the proof. ■

We shall only need Theorem 2 in the case that there is an open neighborhood V of x_0 with $\overline{f(V)}$ compact.

We shall also need a global bifurcation theorem from [52], pp. 106–107. See, also, [50] and [51] for closely related theorems. We refer to Section 4 of [52] for a more detailed discussion of bifurcation theorems in metric ANR's. We collect our assumptions:

H1. X is a metric ANR with metric d, J is an open interval of reals, and $f: X \times J \to X$ is a locally compact map such that $f(x_0,s) = x_0$ for some $x_0 \in X$ and all $s \in J$. Suppose that $A = \{I_k: 1 \leq k \leq m\}$ is a finite or countable collection of pairwise disjoint, compact intervals $I_k \subset J$. We allow I_k to be a point and we write $\Lambda = \bigcup_{k=1}^{m} I_k$. If J_0 is any compact interval, $J_0 \subset J$, assume that $J_0 \cap I_k$ is empty except for finitely many k. If $J_0 \cap I_k$ is empty for all k, suppose that there exists $\varepsilon = \varepsilon(J_0) > 0$ such that $f(x,s) \neq x$ for $0 < d(x,x_0) \leq \varepsilon$ and $s \in J_0$.

If H1 is satisfied and $I_j \in A$, one can define an integer $\Delta(I_j)$ as follows: If $I_j = [c_j,d_j]$, select $\eta>0$ so that $[c_j-\eta, d_j+\eta] \cap I_k$ is empty for all $k \neq j$. Define $\gamma = c_j-\eta$ and $\delta = d_j+\eta$ and select $\varepsilon>0$ so that $f(x,s) \neq x$ for $0 < d(x,x_0) \leq \varepsilon$ and for $s = \gamma$ or $s = \delta$. Let $B_\varepsilon(x_0) = \{x \in X \,|\, d(x,x_0) < \varepsilon\}$, write $f_\lambda(x) := f(x,\lambda)$ and define $\Delta(I_j)$ by

$$\Delta(I_j) = i_X(f_\delta, B_\varepsilon(x_0)) - i_X(f_\gamma, B_\varepsilon(x_0)).$$

With the aid of the basic properties of the fixed point index, it is not hard to prove that $\Delta(I_j)$ is independent of the particular numbers γ, δ and ε chosen above.

Theorem 3. (See Theorem 4.3, p. 107, in [52]). Assume that hypothesis H1 is satisfied and let notation be as in H1. Define

$S = (\{x_0\}\times\Lambda) \cup \{(x,s) \in X\times J: f(x,s) = x$ and $x \neq x_0\}$ and, for fixed $k\geq 1$, define S_k to be the connected component of S which contains $\{x_0\}\times I_k$. Then either (a) S_k is noncompact or (b) S_k is compact, $A_k = \{I \in A: \{x_0\}\times I \subset S_k\}$ is finite and

$$\sum_{I\in A_k} \Delta(I_k) = 0.$$

Theorem 3 is actually true under more general assumptions, e.g., if $X \in \mathcal{F}$ and $f: X\times J \to X$ is a local strict–set–contraction. The proof is the same.

We shall actually only need a very special case of Theorem 3.

Corollary 2. Assume that hypothesis H1 is satisfied and that, in the notation of H1, A contains only one interval I_1. Assume that $\Delta(I_1) \neq 0$. If S and S_1 are as defined in Theorem 3, S_1 is not compact.

Proof. In the notation of Theorem 3,

$$\sum_{I\in A_1} \Delta(I_k) = \Delta(I_1) \neq 0,$$

so case (a) of Theorem 3 must hold. ■

In order to apply Corollary 2, one needs conditions under which one can verify that $\Delta(I_1) \neq 0$.

The following result, which uses Theorems 1 and 2, will suffice for our purposes.

Corollary 3. Let G be a closed, convex infinite dimensional subset of a Banach space E, $0 \in G$, and assume that $f: G\times(0,\infty) \to G$ is a locally compact map with $f(0,s) = 0$ for all $s>0$. Let $I_1 = [c_1,d_1]$ be a compact interval in $(0,\infty)$. If J_0 is any compact interval in $(0,\infty)$ such that $J_0 \cap I_1$ is empty, assume that there exists $\varepsilon = \varepsilon(J_0) > 0$ with $f(x,s) \neq x$ for $0<\|x\|\leq\varepsilon$ and $s \in J_0$. For some λ, $0<\lambda<c_1$, assume that there exists $\varepsilon_1>0$ and a locally compact map $h: G\times[0,1] \to G$ such that (a) $h(x,t) \neq x$ for $0\leq t\leq 1$ and $0<\|x\|\leq\varepsilon_1$, (b) $h(x,0) = f(x,\lambda)$ for $0\leq\|x\|\leq\varepsilon_1$ and (c) the map $x \to h(x,1)$ has 0 as an attractive fixed point. For some μ, $\mu>d_1$, assume that there exists $\eta_1>0$ and a locally compact map $j: G\times[0,1] \to G$ such that

(α) $j(x,t) \neq x$ for $0<\|x\|\leq\eta_1$ and $0\leq t\leq 1$, (β) $j(x,0) = f(x,\mu)$ for $0\leq\|x\|\leq\eta_1$ and (γ) the map $x \longrightarrow j(x,1)$ has 0 as an ejective fixed point. Then, in the notation of Theorem 3, $\Delta(I_1) = -1$ and S_1 is not compact. If $f\colon G\times(0,\infty) \longrightarrow G$ is a k–set–contraction, $k<1$, where $\|(x,t)\| := \|x\| + |t|$ for $(x,t) \in E\times(0,\infty)$), then either (1) there exists a sequence $(x_k,t_k) \in S_1$ with $\lim_{k\to\infty} t_k = 0$ or (2) S_1 is unbounded.

Proof. If $h_1(x) = h(x,1)$, $j_1(x) = j(x,1)$, $f_s(x) := f(x,s)$ and B_ρ denotes $\{x \in G \mid \|x\| < \rho\}$, the homotopy property implies that

$$i_G(f_\lambda,B_{\varepsilon_1}) = i_G(h_1,B_{\varepsilon_1}),$$

and Theorem 1 implies that

$$i_G(h_1,B_{\varepsilon_1}) = 1.$$

A similar argument, using Theorem 2, shows that

$$i_G(f_\mu,B_{\eta_1}) = 0, \quad \text{so}$$

$$\Delta(I_1) = i_G(f_\mu,B_{\eta_1}) - i_G(f_\lambda,B_{\varepsilon_1}) = -1.$$

Corollary 2 now implies that S_1 is not compact. It is easy to see that S_1 is closed in the topology on $G\times(0,\infty)$, so, if there does not exist a sequence $(x_k,t_k) \in S_1$ with $\lim_{k\to\infty} t_k = 0$, S_1 is closed in $E\times\mathbb{R}$. If S_1 is bounded, there exists a compact interval $J_1 \subset (0,\infty)$ with $S_1 \subset f(S_1)\times J_1$. If $E\times\mathbb{R}$ is given the norm $\|(x,t)\| = \|x\| + |t|$ and $\tilde{\gamma}$ is the Kuratowski measure of noncompactness with respect to this norm and γ the corresponding measure of noncompactness with respect to $\|\cdot\|$ on E, our assumption is that for any bounded set $T \subset G\times(0,\infty)$,

$$\gamma(f(T)) \leq k\,\tilde{\gamma}(T).$$

However, if T_1 is any bounded set in E and J_1 any compact interval, one can check that

$$\gamma(T_1) = \tilde{\gamma}(T_1\times J_1).$$

It follows that

$$\tilde{\gamma}(S_1) \leq \tilde{\gamma}(f(S_1)\times J_1) = \gamma(f(S_1)) \leq k\tilde{\gamma}(S_1),$$

so $\tilde{\gamma}(S_1) = 0$ and S_1 has compact closure in $E\times\mathbb{R}$. This implies that S_1 is compact, a contradiction. ∎

An examination of the proof shows that we only need to know that for each $a>0$, $f|G\times[a,\infty)$ is a k–set–contraction for some $k = k_a<1$. Actually, however, we shall only need Corollary 3 in the case that f maps bounded subsets of $G\times(0,\infty)$ to sets with compact closure in G.

We now return to equation $(1)_\lambda$. We shall always assume at least the following about the function g in $(1)_\lambda$.

H2. $g\colon \mathbb{R} \to \mathbb{R}$ is a continuous map such that $ug(u) > 0$ for all $u\neq 0$. The left hand and right hand derivatives of g at 0 both exist and are finite:

$$\lim_{u\to 0^+}\left[\frac{g(u)}{u}\right] = a \quad\text{and}\quad \lim_{u\to 0^-}\left[\frac{g(u)}{u}\right] = b.$$

Under hypothesis H2 we are interested in finding so–called "slowly oscillating periodic solutions" of $(1)_\lambda$. A periodic solution $x(t)$ of $(1)_\lambda$ will be called a slowly oscillating periodic solution if there exist positive real numbers z_0, z_1 and z_2 with $z_0 + 1<z_1$, $z_1 + 1<z_2$, $x(t)>0$ for $z_0<t<z_1$, $x(t)<0$ for $z_1<t<z_2$ and $x(t+z_2-z_0) = x(t)$ for all t. The word "slowly" in the definition refers to the fact that the zeros of $x(t)$ are separated by a distance greater than the time lag 1.

There is by now a standard method for reducing the problem of existence of slowly oscillating periodic solutions of $(1)_\lambda$ to a fixed point problem. See Section 2 of [44], Section 2 of [46] or [48]. The arguments involved here are elementary, but they may fail for more complicated differential–delay equations and must be carefully verified. Let $C[0,1] = E$ denote the Banach space of continuous, real–valued functions on $[0,1]$, with $\|x\| = \max_{0\le t\le 1} |x(t)|$ for $x\in E$. If $x\in E$, we say that x is "increasing on $[0,1]$" if $x(t_1) \le x(t_2)$ whenever $0\le t_1\le t_2\le 1$. Similarly, x is "strictly increasing on $[0,1]$" if $x(t_1)<x(t_2)$ whenever $0\le t_1<t_2\le 1$. "Decreasing" and "strictly decreasing" are defined analogously. Define sets $K, K_1 \subset E$ by

$$\begin{aligned} K &= \{\theta \in C[0,1]\colon \theta(0) = 0 \text{ and } \theta \text{ is increasing}\} \text{ and} \\ K_1 &= \{\theta \in C[0,1]\colon \theta(0) = 0 \text{ and } \theta \text{ is decreasing}\}. \end{aligned} \tag{2}$$

One can check that K and K_1 are closed, convex and infinite dimensional subsets of E.

If a and b are as in H2, define a function $h: \mathbb{R} \to \mathbb{R}$ by

$$h(u) = \begin{cases} au & \text{for } u \geq 0 \\ bu & \text{for } u < 0 \end{cases} \tag{3}$$

and define $j: \mathbb{R}\times[0,1] \to \mathbb{R}$ by

$$j(u,s) = (1-s)g(u) + sh(u). \tag{4}$$

It will be convenient to study a generalization of $(1)_\lambda$. If $\theta \in E$, $\lambda > 0$ and $0 \leq s \leq 1$, there is a unique continuous function $x(t) = x(t; \theta,\lambda,s)$, defined for $t \geq 0$, such that $x|[1,\infty)$ is C^1 and

$$x'(t) = -\lambda j(x(t-1),s) \quad \text{for } t \geq 1 \text{ and} \tag{5}$$
$$x|[0,1] = \theta.$$

If $\theta \in K-\{0\}$, define $z_1 = z_1(\theta,\lambda,s)$ to be the first time $t > 1$ such that $x(t; \theta,\lambda,s) = 0$, if such a t exists. Otherwise, define $z_1(\theta,\lambda,s) = \infty$. In general, if $z_m = z_m(\theta,\lambda,s)$ is defined and finite, define $z_{m+1}(\theta,\lambda,s)$ to be the first time $t > z_m(\theta,\lambda,s)$ such that $x(t; \theta,\lambda) = 0$. If $z_1(\theta,\lambda,s) < \infty$ for $\theta \in K-\{0\}$, one can easily prove (see Section 2 of [44] or [46] or [48]) that $x(t; \theta,\lambda,s) > 0$ for $z_1 - 1 \leq t < z_1$ and $x'(t) < 0$ for $z_1 \leq t < z_1 + 1$. Thus $z_2 - z_1 > 1$ and generally $z_{m+1} - z_m > 1$ if $z_m < \infty$. Define a map $R: K\times(0,\infty)\times[0,1] \to K_1$ by $R(\theta,\lambda,s) = 0$ if $\theta = 0$ or $z_1(\theta,\lambda,s) = \infty$ and $R(\theta,\lambda,s) = \psi$, if $z_1(\theta,\lambda,s) < \infty$, where

$$\psi(t) = x(z_1(\theta,\lambda,s) + t; \theta,\lambda,s) \quad \text{for } 0 \leq t \leq 1.$$

One can prove (see Section 2 of [46], [44], [45] or [48]) that the map R is continuous and takes bounded subsets of $K\times(0,\infty)\times[0,1]$ into precompact sets (i.e, sets with compact closure). Similarly, if $\theta \in K_1-\{0\}$, define $\zeta_1 = \zeta_1(\theta,\lambda,s)$ to be the first $t > 1$ such that $x(t; \theta,\lambda,s) = 0$, if such a t exists. Otherwise, define $\zeta_1(\theta,\lambda,s) = \infty$. Define $S: K_1\times(0,\infty)\times[0,1] \to K$ by $S(\theta,\lambda,s) = \psi$, where $\psi = 0$ if $\zeta_1(\theta,\lambda,s) = \infty$ or $\theta = 0$ and otherwise

$$\psi(t) = x(\zeta_1(\theta,\lambda,s) + t; \theta,\lambda,s), \quad 0 \leq t \leq 1.$$

S is also a continuous map and takes bounded sets to precompact sets. If we define

$$\Phi: K\times(0,\infty)\times[0,1] \to K \quad \text{by}$$

$$\Phi(\theta,\lambda,s) = S(R(\theta,\lambda,s),\lambda,s), \tag{6}$$

it is well–known that equation (5) has a slowly oscillating periodic solution $x(t)$ for a given $\lambda>0$ and $s \in [0,1]$ if and only if there exists $\theta \in K-\{0\}$ with

$$\Phi(\theta,\lambda,s) = \theta.$$

For notational convenience we define

$$G(\theta,\lambda) = \Phi(\theta,\lambda,0) \quad \text{and} \quad H(\theta,\lambda) = \Phi(\theta,\lambda,1). \tag{7}$$

We shall also write

$$G_\lambda(\theta) := G(\theta,\lambda) \quad \text{and} \quad H_\lambda(\theta) := H(\theta,\lambda). \tag{8}$$

The key to completing our analysis of equation $(1)_\lambda$ is a careful analysis of equation (5) for $s=1$. Our arguments will have much in common with those in [47].

Lemma 3. Assume that $\theta \in K$ or $\theta \in K_1$ (see equation (2)) and define $h: \mathbb{R} \to \mathbb{R}$ by $h(u) = \alpha u$ for $u \geq 0$ and $h(u) = \beta u$ for $u<0$, where $\alpha>0$ and $\beta>0$. Let $x(t)$ be the solution of

$$\begin{aligned} x'(t) &= -h(x(t-1)), \quad t \geq 1 \\ x|[0,1] &= \theta. \end{aligned} \tag{9}$$

If $\theta \in K-\{0\}$, let z_1 denote the first $t>1$ such that $x(t) = 0$ (if such a t exists); and if $\theta \in K_1-\{0\}$, let ζ_1 denote the first $t>1$ such that $t=0$ (if such a t exists). Then it follows that if $\theta \in K$,

$$|x(z_1+1)| \leq [\max(\alpha - \tfrac{1}{2}, \tfrac{\alpha}{2})]\, x(1). \tag{10}$$

If $\theta \in K_1$, it follows that

$$|x(\zeta_1+1)| \leq [\max(\beta - \tfrac{1}{2}, \tfrac{\beta}{2})]\, |x(1)|. \tag{11}$$

Proof. We assume that $\theta \in K$, since the proof is exactly analogous if $\theta \in K_1$. We remark that if $\alpha \geq 1$, it is trivial to see that $z_1 \leq 3$: If $x(t) > 0$ on $[1,2]$, then (because $x(t)$ is decreasing on $[1,2]$) we have that $x'(t) = -\alpha x(t-1) \leq -\alpha x(2)$ for $2 \leq t \leq 3$, which implies that $x(t) = 0$ for some $t \in [2, 2 + \alpha^{-1}]$. A subtler argument actually shows that $z_1<\infty$ if $\alpha>e^{-1}$.

We prove inequality (10) by considering several subcases.

<u>Case 1</u>. $1<z_1\le 2$. We see from (9) that x is decreasing on $[1,2]$ and concave down on $[1,2]$. It follows that

$$x(t) \le x'(z_1)(t-z_1) = \alpha(z_1-t)x(z_1-1) \qquad (12)$$

for $1\le t\le z_1$. Because x is increasing on $[0,1]$ we have that

$$x(1) = -\int_1^{z_1} x'(t)dt = \alpha\int_0^{z_1-1} x(t)dt < \alpha(z_1-1)x(z_1-1).$$

It follows that there exists τ, $1<\tau<z_1$, with

$$\alpha(z_1-\tau)x(z_1-1) = x(1).$$

For $0\le t\le \tau$ we have the estimate

$$x(t) \le x(1). \qquad (13)$$

Using (13) we see that

$$x(1+\tau) = -\alpha\int_{z_1-1}^{\tau} x(t)dt \ge -\alpha(1+\tau-z_1)x(1)$$

$$\ge -\alpha\left[1 - \frac{x(1)}{\alpha x(z_1-1)}\right] x(1).$$

Similarly, we obtain from (12) that

$$x(z_1+1) - x(1+\tau) = -\alpha\int_{\tau}^{z_1} x(t)dt$$

$$\ge -\alpha\int_{\tau}^{z_1} \alpha x(z_1-1)(z_1-t)dt = -\frac{(x(1))^2}{2x(z_1-1)}.$$

Combining these two inequalities, we find that

$$|x(z_1+1)| \le \left[\alpha - \left[\frac{1}{2}\right]\left[\frac{x(1)}{x(z_1-1)}\right]\right] x(1) \le \left[\alpha - \left[\frac{1}{2}\right]\right] x(1).$$

<u>Case 2</u>. $2<z_1\le 3$. The function x is concave down and decreasing on $[1,2]$ and $x(2)>0$. It follows as in Case 1 that

$$x(t) \le \alpha(2-t)x(1) + x(2)$$

for $1\le t\le 2$. Thus there exists τ, $1\le\tau\le 2$, with

$$\alpha(2-\tau)x(1) + x(2) = x(1).$$

For $0 \leq t \leq \tau$, we again have the estimate $x(t) \leq x(1)$.

A calculation gives that

$$2 - \tau = [x(1) - x(2)][\alpha x(1)]^{-1}.$$

Because $x'(t) \geq -\alpha x(1)$ on $[2, z_1]$, we see that

$$z_1 - 2 \geq x(2)[\alpha x(1)]^{-1},$$

so we conclude that

$$z_1 - \tau \geq \alpha^{-1}.$$

The function x is convex on $[2, z_1+1]$. Thus, if $z_1 - 1 \geq \tau$, one can see that

$$x(z_1+1) = -\alpha \int_0^1 x(z_1-1+t)dt \geq -\alpha \int_0^1 (1-t)x(1)dt = -\frac{\alpha}{2} x(1),$$

and we find in this case that

$$|x(z_1+1)| \leq \left[\frac{\alpha}{2}\right] x(1).$$

If $z_1 - 1 < \tau$ we have

$$x(1+\tau) = -\alpha \int_{z_1-1}^{\tau} x(t)dt \geq -\alpha(\tau+1-z_1)x(1).$$

On the interval $[\tau, z_1]$, we again see that $x(t)$ is less than or equal to a convex function $y(t)$ with $y(\tau) = x(1)$ and $y(z_1) = 0$, so we deduce that

$$x(z_1+1) - x(\tau+1) = -\alpha \int_{\tau}^{z_1} x(t)dt \geq -\alpha \int_{\tau}^{z_1} (z_1-t)(z_1-\tau)^{-1} x(1)dt$$

$$\geq -\left[\frac{1}{2}\right] \alpha(z_1-\tau)\, x(1).$$

Combining these inequalities we find

$$x(z_1+1) \geq \left[-\alpha + \left[\frac{1}{2}\right] \alpha(z_1-\tau)\right] x(1).$$

Because we have observed that $(z_1-\tau) \geq \alpha^{-1}$ we conclude that

$$|x(z_1+1)| \leq \left[\alpha - \left[\frac{1}{2}\right]\right] x(1).$$

Thus, in case 2, we always have

$$|x(z_1+1)| \leq \max\left[\alpha - \left[\frac{1}{2}\right], \left[\frac{\alpha}{2}\right]\right] x(1).$$

Case 3. $z_1>3$. The function x is convex and decreasing on $[z_1-1,z_1]$, so

$$x(z_1-1+t) \leq (1-t)x(z_1-1) \leq (1-t)\ x(1), \quad 0\leq t\leq 1.$$

It follows that

$$|x(z_1+1)| = \alpha\int_0^1 x(z_1-1+t)dt \leq \left[\frac{\alpha}{2}\right] x(1). \qquad \blacksquare$$

As an immediate consequence of Lemma 3 we obtain

Lemma 4. Suppose that $h\colon \mathbb{R} \to \mathbb{R}$ is given by equation (3), where $a>0$ and $b>0$, and that $H(\theta,\lambda)$ is defined as in (7) for $\theta \in K$, $\lambda>0$. Then it follows that for $\lambda>0$ and $\theta \in K$

$$\|H(\theta,\lambda)\| \leq k_\lambda \|\theta\|,$$

$$k_\lambda = \left[\max\left[\lambda a - \frac{1}{2}, \frac{\lambda a}{2}\right]\right]\left[\max\left[\lambda b - \frac{1}{2}, \frac{\lambda b}{2}\right]\right].$$

In particular, if $\lambda \max(a,b) < \left[\frac{3}{2}\right]$, 0 is an asymptotically stable fixed point of $H_\lambda\colon K \to K$.

It remains to investigate when 0 is an ejective fixed point of H_λ. The reader should compare arguments in [47].

Lemma 5. Let hypotheses and notation be as in Lemma 3. If $\alpha\geq 2$ and $\theta \in K-\{0\}$, it follows that

$$|x(z_1+1)| \geq (\sqrt{2\alpha} - 1)\ x(1).$$

If $1\leq\alpha\leq 2$, $|x(z_1+1)| \geq \left[\left[\frac{\alpha}{2}\right] - \sqrt{\left[\frac{2-\alpha}{3}\right]}\left[\frac{2-\alpha}{3}\right]\right] x(1)$.

If $\beta\geq 2$ and $\theta \in K_1-\{0\}$, it follows that

$$|x(\zeta_1+1)| \geq (\sqrt{2\beta} - 1)\ |x(1)|.$$

If $1\leq\beta\leq 2$, $|x(\zeta_1+1)| \geq \left[\left[\frac{\beta}{2}\right] - \sqrt{\left[\frac{2-\beta}{3}\right]}\left[\frac{2-\beta}{3}\right]\right] |x(1)|$.

Proof. We only consider the case $\theta \in K-\{0\}$; the proof for $\theta \in K_1-\{0\}$ is analogous. As was remarked in the proof of Lemma 3, we have $z_1\leq 3$ because $\alpha\geq 1$. We consider two cases.

Case 1. $1<z_1\leq 2$. Integrating equation (9) from 1 to z_1 gives

$$x(1) = \alpha\int_0^{z_1-1} x(t)dt \leq \alpha(z_1-1)x(z_1-1).$$

It follows that

$$|x(2)| = \alpha\int_{z_1-1}^{1} x(t)dt \geq \alpha(2-z_1)x(z_1-1)$$

$$\geq \alpha(2-z_1)\left[\frac{x(1)}{\alpha(z_1-1)}\right] = \left[\frac{2-z_1}{z_1-1}\right] x(1).$$

Integrating (9) from 2 to z_1+1 gives

$$|x(z_1+1) - x(2)| = \alpha\int_1^{z_1} x(t)dt.$$

We know that x is concave down on $[1,2]$, so $x(t) \geq (z_1-t)(z_1-1)^{-1}x(1)$ for $1\leq t\leq 2$; and we obtain

$$|x(z_1+1) - x(2)| \geq \alpha\int_1^{z_1} (z_1-t)(z_1-1)^{-1}x(1)dt$$

$$= \left[\frac{\alpha}{2}\right](z_1-1)x(1).$$

Combining these inequalities gives

$$|x(z_1+1)| = |x(2)| + |x(z_1+1) - x(2)|$$

$$\geq \left[\left[\frac{\alpha}{2}\right](z_1-1) + \left[\frac{2-z_1}{z_1-1}\right]\right] x(1).$$

It is a simple calculus exercise that

$$\min_{1<z_1\leq 2}\left[\left[\frac{\alpha}{2}\right](z_1-1) + \left[\frac{2-z_1}{z_1-1}\right]\right] = \begin{cases} \sqrt{2\alpha} - 1, & \text{if } \alpha\geq 2 \\ \left[\frac{\alpha}{2}\right], & \text{if } 1\leq\alpha\leq 2. \end{cases}$$

Thus, in case 1, we conclude that

$$|x(z_1+1)| \geq \begin{cases} (\sqrt{2\alpha} - 1)\, x(1) & \text{if } \alpha\geq 2 \\ \left[\frac{\alpha}{2}\right] x(1), & \text{if } 1\leq\alpha\leq 2. \end{cases}$$

Case 2. $2<z_1\leq 3$. We know that x is concave down on $[1,2]$ and convex

on $[2,z_1+1]$. It follows that

$$x(t) \geq (2-t)x(1) + (t-1)x(2) \quad \text{for} \quad 1\leq t\leq 2,$$

and we deduce that

$$x(3) = x(2) - \alpha\int_1^2 x(t)dt \leq x(2) - \alpha\int_1^2 [(2-t)x(1) + (t-1)x(2)]dt$$

$$\leq x(2) - \left[\frac{\alpha}{2}\right][x(1) + x(2)].$$

If $\alpha\geq 2$, we deduce that

$$x(z_1+1) \leq x(3) \leq -\left[\frac{\alpha}{2}\right]x(1) \quad \text{and}$$

$$\left[\frac{\alpha}{2}\right]x(1) \leq |x(z+1)|.$$

Thus we can assume that $1\leq\alpha<2$. Because x is convex on $[2,z_1]$, we have that

$$x(t) \geq x'(z_1)(t-z_1) = \alpha x(z_1-1)(z_1-t) \quad \text{for} \quad 2\leq t\leq z_1.$$

Using this fact and the above estimate for $x(3)$ we obtain

$$x(z_1+1) = x(3) + (x(z_1+1)-x(3)) \leq x(2) - \left[\frac{\alpha}{2}\right][x(1)+x(2)] - \alpha\int_2^{z_1} x(t)dt$$

$$\leq x(2) - \left[\frac{\alpha}{2}\right][x(1)+x(2)] - \left[\frac{\alpha^2(z_1-2)^2x(z_1-1)}{2}\right].$$

Because $x'(t) \leq -\alpha x(1)$ for $2\leq t\leq z_1$, we have that $x(2) - \alpha(z_1-2)x(1) \leq 0$ and $x(2)x(1)^{-1} \leq \alpha(z_1-2)$. If we use this lower bound for $\alpha(z_1-2)$ and define $\rho = x(2)x(1)^{-1}$, so $0<\rho\leq 1$, we obtain

$$x(z_1+1) \leq x(1)\left[-\left[\frac{\alpha}{2}\right] + \rho - \left[\frac{\alpha}{2}\right]\rho - \left[\frac{1}{2}\right]\rho^2x(z_1-1)(x(1))^{-1}\right]$$

$$\leq x(1)\left[-\left[\frac{\alpha}{2}\right] + \rho - \left[\frac{\alpha}{2}\right]\rho - \left[\frac{1}{2}\right]\rho^3\right].$$

An easy calculus exercise shows that

$$\max_{0<\rho\leq 1}\left[-\left[\frac{\alpha}{2}\right] + \rho - \left[\frac{\alpha}{2}\right]\rho - \left[\frac{1}{2}\right]\rho^3\right] = -\frac{\alpha}{2} + \sqrt{\left[\frac{2-\alpha}{3}\right]}\left[\frac{2-\alpha}{3}\right].$$

We conclude that in case 2, if $1\leq\alpha\leq 2$,

$$|x(z_1+1)| \geq \left[\frac{\alpha}{2} - \sqrt{\left[\frac{2-\alpha}{3}\right]\left[\frac{2-\alpha}{3}\right]}\right] x(1). \qquad \blacksquare$$

Lemma 5 immediately yields conditions under which 0 is an ejective fixed point of H_λ.

Lemma 6. Suppose that $a,b>0$, $h: \mathbb{R} \to \mathbb{R}$ is given by equation (3) and $H: K\times(0,\infty) \to K$ is given by equation (7). For $r\geq 1$, define a function $\mu(r)$ by

$$\mu(r) = \begin{cases} \sqrt{2r} - 1, & \text{for } r\geq 2 \\ \left[\frac{r}{2}\right] - \sqrt{\left[\frac{2-r}{3}\right]\left[\frac{2-r}{3}\right]}, & \text{for } 1\leq r\leq 2. \end{cases}$$

If $\theta \in K$, $\lambda>0$ and $\min(\lambda a,\lambda b) \geq 1$, then

$$\|H(\theta,\lambda)\| \geq \mu(\lambda a)\mu(\lambda b)\|\theta\|.$$

In particular, if $\lambda \min(a,b) > 2$, then $\mu(\lambda a)\mu(\lambda b) = (\sqrt{2\lambda a} - 1)(\sqrt{2\lambda b} - 1) > 1$ and 0 is an ejective fixed point of H_λ.

Obviously, by using Lemma 6 more carefully, we could describe situations in which $1\leq\lambda a\leq 2\leq\lambda b$ or $1\leq\lambda b\leq 2\leq\lambda a$ and 0 is an ejective fixed point of H_λ.

With the aid of Lemmas 4 and 6 we can now make the crucial step in verifying the hypotheses of Corollary 3.

Lemma 7. Assume that $g: \mathbb{R} \to \mathbb{R}$ satisfies hypothesis H2 and that $a>0$ and $b>0$. Let $\Phi: K\times(0,\infty)\times[0,1] \to K$ be defined by equation (6). Define numbers c_1 and d_1 by

$$c_1 \max(a,b) = 3/2 \quad \text{and} \quad d_1 \min(a,b) = 2.$$

Let J_0 be a compact interval, $J_0 \subset (0,\infty)$, such that $J_0 \cap [c_1,d_1]$ is empty. Then there exists $\varepsilon = \varepsilon(J_0) > 0$ such that

$$\Phi(\theta,\lambda,s) \neq \theta$$

for $0<\|\theta\|\leq\varepsilon$, $\lambda \in J_0$ and $0\leq s\leq 1$. If $\lambda \notin [c_1,d_1]$ and $\varepsilon>0$ is such that $H_\lambda(\theta) \neq \theta$ and $G_\lambda(\theta) \neq \theta$ for $0<\|\theta\|\leq\varepsilon$, then $i_K(H_\lambda,B_\varepsilon) = i_K(G_\lambda,B_\varepsilon) = 0$ for $\lambda>d_1$ and $i_K(H_\lambda,B_\varepsilon) = i_K(G_\lambda,B_\varepsilon) = 1$ for $0<\lambda<c_1$. (Here $B_\varepsilon = \{\theta\in K: \|\theta\|<\varepsilon\}$).

Proof. If we can prove the first part of Lemma 7, the homotopy property for the fixed point index implies that for $\lambda \notin [c_1,d_1]$ and $\varepsilon>0$ sufficiently small,

$$i_K(H_\lambda,B_\varepsilon) = i_K(G_\lambda,B_\varepsilon).$$

Lemma 4 implies that 0 is an asymptotically stable fixed point of H_λ for $\lambda>c_1$, and Theorem 1 then implies that $i_K(H_\lambda,B_\varepsilon) = 1$. Lemma 6 implies that 0 is an ejective fixed point of H_λ for $\lambda>d_1$, and Theorem 2 gives $i_K(H_\lambda,B_\varepsilon) = 0$.

To prove the first part of the lemma, we assume not. Then there exists a sequence (θ_k,λ_k,s_k), $k\geq 1$, with

$$\|\theta_k\| > 0, \quad \lim_{k\to\infty} \|\theta_k\| = 0, \quad \lambda_k \in J_0, \quad 0\leq s_k\leq 1, \quad \text{and}$$

$$\Phi(\theta_k,\lambda_k,s_k) = \theta_k.$$

By taking subsequences we can assume that $\lambda_k \longrightarrow \lambda \in J_0$ and $s_k \longrightarrow s \in [0,1]$. By construction, there exists a slowly oscillating periodic solution x_k, defined on $\mathbb{R}$, with

$$x_k'(t) = -\lambda_k j(x_k(t-1),s_k)$$

$$x_k|[0,1] = \theta_k,$$

where j is defined by (4). We leave to the reader the easy verification that there exists a constant M such that

$$\|x_k\|: = \sup_t |x_k(t)| \leq M\|\theta_k\|,$$

so $\|x_k\| \longrightarrow 0$. If we define $y_k(t) = x_k(t)\|x_k\|^{-1}$, we find by integrating the differential equation for $x_k(t)$ that

$$y_k(t) - y_k(1) = -\lambda_k \int_1^t [(1-s_k)g(x_k(s-1))\|x_k\|^{-1} + s_k h(y_k(s-1))]ds.$$

By using H2, one can see that

$$\lim_{k\to\infty} [g(x_k(s-1))\|x_k\|^{-1} - h(y_k(s-1))] = 0,$$

uniformly in $s \in \mathbb{R}$. It then follows that $y_k'(t)$ is uniformly bounded in t and k. Since $\|y_k\| = 1$, the Ascoli–Arzela theorem implies that, by taking a subsequence, we can assume that $y_k(t)$ converges to a continuous function $y(t)$, uniformly on compact intervals in $\mathbb{R}$.

There is one subtle point to consider: If z_{k1} denotes the first $t>1$ such that $x_k(t) = 0$ and z_{k2} denotes the first $t>z_{k1}$ with $x_k(t) = 0$, one must prove that z_{k2}, the period of x_k, is bounded uniformly in k. However, if z_{k2} is not uniformly bounded, one can prove that

$$x_k(z_{k2}+1) \leq \frac{1}{2} x_k(1)$$

for large k, which contradicts the definition of a slowly oscillating periodic solution. We omit further details.

Once one knows that z_{2k} is bounded, it follows easily that $\|y\| = 1$ and y is a slowly oscillating periodic solution of

$$y(t) - y(1) = -\lambda \int_1^t h(y(s-1))ds.$$

This implies that y satisfies the equation

$$y'(t) = - \lambda h(y(t-1)).$$

But Lemmas 4 and 6 imply that, because $\lambda \notin [c_1,d_1]$, no such solution exists, so we have obtained a contraction. ∎

We can now give our first existence result for slowly oscillating periodic solutions of equation $(1)_\lambda$.

Theorem 4. Assume that g satisfies hypothesis H2 with $a>0$ and $b>0$. Define numbers c_1 and d_1 by

$$c_1 \max(a,b) = \left[\frac{3}{2}\right] \quad \text{and} \quad d_1 \min(a,b) = 2. \qquad (14)$$

Let $G: K\times(0,\infty) \to K$ be defined by equation (7). Define a set S by

$$S = (\{0\}\times[c_1,d_1]) \cup \{(\theta,\lambda) \in K\times(0,\infty): \theta\neq0 \text{ and } G(\theta,\lambda) = \theta\}.$$

Let S_1 be the connected component of S which contains $\{0\}\times[c_1,d_1]$. Then S_1 is not a compact subset of $K\times(0,\infty)$, but S_1 is closed as a subset of $K\times(0,\infty)$.

Proof. This is direct consequence of Corollary 2. In the notation of Corollary 2, with $I_1 = [c_1,d_1]$, Lemma 7 implies that $\Delta(I_1) = -1$. Lemma 7 also shows that hypothesis H1 is satisfied. The fact that S_1 is closed as a subset of $K\times(0,\infty)$ is a simple point set topology exercise and is left to the reader. ∎

Corollary 4. Let a and b be positive reals and suppose that $h: \mathbb{R} \to \mathbb{R}$ is defined by $h(u) = au$ for $u \geq 0$ and $h(u) = bu$ for $u<0$. Let positive numbers c_1 and d_1 be defined by equation (14). Then there exists a number λ_1 with $c_1 \leq \lambda_1 \leq d_1$, such that the equation

$$x'(t) = -\lambda_1 h(x(t-1))$$

has a slowly oscillating periodic solution.

Proof. If H is defined by equation (7), Lemmas 4 and 6 imply that the equation $H(\theta,\lambda) = \theta$, $\theta \in K-\{0\}$, $\lambda>0$, has no solution for $0<\lambda<c_1$ or for $\lambda>d_1$. If S is defined as in Theorem 4, i.e., $S = (\{0\}\times[c_1,d_1]) \cup \{(\theta,\lambda) \in K\times(0,\infty)$: $\theta \neq 0$ and $H(\theta,\lambda) = \theta\}$ and if S_1 denotes the connected component of S which contains $\{0\}\times[c_1,d_1]$, Theorem 4 implies that S_1 is closed in $K\times(0,\infty)$ and not compact. It follows that S_1 is not bounded, so there exists $(\theta_1,\lambda_1) \in S_1$, $\theta_1 \neq 0$, and a corresponding slowly oscillating periodic solution $x_1(t)$ with $x_1|[0,1] = \theta_1$ and

$$x_1'(t) = -\lambda_1 h(x_1(t-1)). \qquad \blacksquare$$

Of course, for the particular function h, if x is a slowly oscillating periodic solution of

$$x'(t) = -\lambda h(x(t-1))$$

and $\alpha>0$, then $y(t) = \alpha x(t)$ is also a slowly oscillating periodic solution.

In order to exploit Theorem 4 we need conditions on g which give bounds on the norm of any slowly oscillating periodic solution of equation $(1)_\lambda$. We present here possibly the simplest condition which implies such bounds. We refer to [45] for examples of other hypotheses which imply bounds.

H3. The function $g: \mathbb{R} \to \mathbb{R}$ satisfies H2, so g is continuous, $ug(u) > 0$ for all $u \neq 0$, $\lim_{u\to 0^+} g(u)u^{-1} = a$ and $\lim_{u\to 0^-} g(u)u^{-1} = b$. In addition, there exists $B>0$ with $g(u) > -B$ for all u.

The following lemma is a well-known result (see Section 2 of [44] or Section 2 of [46]), but we include a proof for completeness.

Lemma 8. (See [44], [46]). Let $g: \mathbb{R} \to \mathbb{R}$ be a map which satisfies hypothesis H3 and let $G: K\times(0,\infty) \to K$ be the corresponding map given by equation (7). If $G(\theta_k,\lambda_k) = \theta_k$ for a sequence $(\theta_k\lambda_k) \in K\times(0,\infty)$, $k\geq1$, with $\theta_k\neq0$, and λ_k bounded, then there exists a constant M with $\|\theta_k\| \leq M$ for all $k\geq1$. There exists a constant $\delta>0$ such that $G(\theta,\lambda) \neq \theta$ for $\theta \in K-\{0\}$ and $0<\lambda\leq\delta$.

Proof. Suppose that (θ_k,λ_k) is a sequence as above with $\lambda_k\leq M_1$. Let x_k be a slowly oscillating periodic solution corresponding to (θ_k,λ_k), so

$$x_k'(t) = -\lambda_k g(x_k(t-1)), \quad \text{for all } t$$
$$x_k|[0,1] = \theta_k.$$

Let z_{k1} be the first $t>1$ such that $x_k(t) = 0$ and z_{k2} be the first $t>z_{k1}$ with $x_k(t) = 0$. By definition we have that

$$x_k(z_{k2}+t) = \theta_k(t), \quad 0\leq t\leq1,$$

so $\|\theta_k\| = \theta_k(1) = x_k(z_{k2}+1)$. However, because g is bounded below we have

$$x_k(z_{k2}+1) = -\lambda_k \int_{z_{k2}-1}^{z_{k2}} g(x_k(s))ds \leq \lambda_k B \leq BM_1,$$

which proves the first part of the lemma.

If the second part of the lemma is false, there exists a sequence $(\theta_k,\lambda_k) \in K\times(0,\infty)$, $\theta_k\neq0$, with $G(\theta_k,\lambda_k) = \theta_k$ and $\lim_{k\to\infty} \lambda_k = 0$. If x_k is defined as above, our estimates above show that

$$\|\theta_k\| \leq B\lambda_k \longrightarrow 0.$$

Define a function $M_2(r)$ by $M_2(r) = \sup_{0\leq u\leq r} g(u)$. Note that

$$|x_k(z_{k1}+1)| = \lambda_k \int_{z_{k1}-1}^{z_{k1}} g(x_k(s))ds \leq \lambda_k M_2(\lambda_k B).$$

This implies that $\delta_k = \sup_s |x_k(s)| \longrightarrow 0$ as $k \longrightarrow \infty$. By using H3 we see that there exists $\eta_0>0$ and $C>0$ with

$$|g(u)| \leq C|u| \quad \text{for} \quad |u|\leq\eta.$$

It follows that for all sufficiently large k we have

$$|x_k(z_{k1}+1)| = \lambda_k \int_{z_{k1}-1}^{z_{k1}} g(x_k(s))ds \le C\lambda_k |x_k(1)| \quad \text{and}$$

$$x_k(z_{k2}+1) = \lambda_k \int_{z_{k2}-1}^{z_{k2}} |g(x_k(s))|ds \le C\lambda_k |x_k(z_{k1}+1)|.$$

Since $x_k(z_{k2}+1) = x_k(1)$, this gives

$$x_k(1) \le (C\lambda_k)^2 x_k(1),$$

which is impossible for k so large that $C\lambda_k < 1$. ∎

Theorem 5. Suppose that $g: \mathbb{R} \to \mathbb{R}$ is a continuous map such that $ug(u) > 0$ for all $u \neq 0$, $\lim_{u\to 0^+} g(u)u^{-1} = a>0$ and $\lim_{u\to 0^-} g(u)u^{-1} = b>0$. Assume that there exists $B>0$ with $g(u) > -B$ for all $u \in \mathbb{R}$. Define numbers c_1 and d_1 by

$$c_1 \max(a,b) = \left[\frac{3}{2}\right] \quad \text{and} \quad d_1 \min(a,b) = 2.$$

Then for every $\lambda > d_1$ the equation

$$x'(t) = -\lambda g(x(t-1))$$

has a slowly oscillating periodic solution.

Proof. Let S and S_1 be as defined in Theorem 4, so S_1 is closed in $K\times(0,\infty)$ and not compact. Lemma 8 implies that there exists $\delta>0$ so that if $(\theta,\lambda) \in S$, $\lambda \ge \delta$. Since G takes bounded sets to precompact sets, Corollary 3 implies that S_1 is unbounded. However, Lemma 8 implies that if S_1 is unbounded, there exists a sequence $(\theta_k,\lambda_k) \in S_1$ with $\lim_{k\to\infty} \lambda_k = \infty$.

We claim that for every $\alpha > d_1$ there exists a slowly oscillating periodic solution of

$$x'(t) = -\alpha\, g(x(t-1)).$$

If, for some $\alpha > d_1$, this is false, then there does not exist a point $(\theta,\alpha) \in S_1$. Let $U = \{(\theta,\lambda) \in S_1: \lambda<\alpha\}$ and $V = (\theta,\lambda) \in S_1: \lambda>\alpha\}$, so $S_1 = U \cup V$ and U and V are open subsets of S_1 in the relative topology on S_1. Since V contains (θ_k,λ_k) for $\lambda_k > \alpha$, V is nonempty. On the other hand, by the definition of S_1, U contains $\{0\}\times[c_1,d_1]$, and hence U is nonempty. We have written S_1 as the union of two disjoint, nonempty open sets, which contradicts the connectedness of S_1. ■

Remark 3. Actually Theorem 4 provides more information than is stated in Theorem 5: In the precise sense given Theorem 4, there is an unbounded, connected set of slowly oscillating periodic solutions of equation $(1)_\lambda$. As is shown in [46] and [35], connectedness can be exploited to give branches of "rapidly oscillating" periodic solutions of $(1)_\lambda$.

Remark 4. In Theorem 5 we have estimated the number d_1 crudely. If we use Lemmas 3–6 more carefully, we find that it suffices to choose d_1 so that 0 is an ejective fixed point of H_λ for $\lambda > d_1$. An examination of Lemma 6 shows that if $b > a$ it suffices to take d_1 to be the infimum of numbers λ such that $\max(2b^{-1}, a^{-1}) \leq \lambda \leq 2a^{-1}$ and

$$(\sqrt{2\lambda b} - 1)\left(\left[\frac{\lambda a}{2}\right] - \sqrt{\left[\frac{2-\lambda a}{3}\right]}\left[\frac{2-\lambda a}{3}\right]\right) \geq 1.$$

If $a > b$, we reverse the roles of a and b in these inequalities. If g satisfies H3 and $\lambda > d_1$ for d_1 chosen in this way, then equation $(1)_\lambda$ has a slowly oscillating periodic solution. Using this observation one can see that if $a = 1$ and $b = 4$ in hypothesis H3 and g satisfies H3, then equation $(1)_\lambda$ has a slowly oscillating periodic solution for $\lambda \geq 1.22$. This contrasts with Theorem 5, which only guarantees existence of slowly oscillating periodic solutions for $\lambda > 2$.

Acknowledgements: The author's work on this paper was partially supported by NSFDMS 9105930.

References

[1] R. Arens and J. Eells, On embedding uniform and topological spaces, Pacific J. Math 6 (1956), 397–403.

[2] R.H. Bing, The elusive fixed point property, American Math. Monthly, February, (1969), 119–132.

[3] K. Borsuk, Theory of Retracts, Polish Sci. Publ., Warsaw, 1967.

[4] F.E. Browder, On a generalization of the Schauder fixed point theorem, Duke Math. J. 26 (1959), 291–304.

[5] ________________, Another generalization of the Schauder fixed point theorem, Duke Math. J. 32 (1965), 399–406.

[6] ______________, A further generalization of the Schauder fixed point theorem, Duke Math. J. 32 (1965), 575–578.

[7] ______________, On the fixed point index for continuous mappings of locally connected spaces, Summa Brasil. Math. 4 (1960), 253–293.

[8] ______________, Local and global properties of nonlinear mappings in Banach spaces, Instituto Naz. di Alta Mat. Symposia Math. 2 (1968), 13–35.

[9] ______________, Asymptotic fixed point theorems, Math. Ann. 185 (1970), 38–60.

[10] R.F. Brown, The Lefschetz Fixed Point Theorem, Scott Foresman Co., Glenview, Illinois, 1971.

[11] G. Darbo, Punti uniti in trasformazioni a condiminio non compatto, Rend. Sem. Math. Univ. Padova 24 (1955), 84–92.

[12] A. Deleanu, Théorie des points fixes sur les retractes de voisinage des espaces convexoides, Bull. Soc. Math. France 89 (1959), 235–243.

[13] A. Dold, Fixed point index and fixed point theorems for euclidean neighborhood retracts, Topology 4 (1965), 1–8.

[14] ______________, Lectures on Algebraic Topology, vol. 200, Springer–Verlag, New York, 1972.

[15] J. Dugundji, An extension of Tietze's theorem, Pacific J. Math. 1 (1951), 353–367.

[16] J. Dugundji and A. Granas, Fixed Point Theory, vol. 1, Polska Akademia Nauk, Monografie Matematyczne, Tom 61, Warsaw, 1982.

[17] J. Eells and G. Fournier, La théorie des points fixes des applications à itérée condensante, Bull Soc. Math. France, Mémoire 45 (1976), 91–120.

[18] C. Fenske and H.–O. Peitgen, Fixed points of zero index in asymptotic fixed point theory, Pacific J. Math 66 (1976), 391–410.

[19] G. Fournier, Généralisations du théoreme de Lefschetz pour des espaces non–compacts, II. Bull. Acad. Pol. Sci. 23 (1975), 701–706.

[20] R.L. Frum–Ketkov, Mappings into a Banach space sphere, Dokl. Akad. Nauk. SSSR 175 (1967), 1229–1231 = Soviet Math. Dokl. 8 (1967), 1004–1007.

[21] R.B. Grafton, A periodicity theorem for autonomous functional differential equations, J. Differential Equations 6 (1969), 87–109.

[22] A. Granas, Some theorems in fixed point theory, Bull Acad. Polon. Sci. 17 (1969), 131–137.

[23] ______________, The Leray–Schauder index and the fixed point theory for arbitrary ANR's, Bull. Soc. Math. France 100 (1972), 209–228.

[24] J. Hale, Asymptotic Behaviour of Dissipative Systems, Mathematical Surveys and Monographs, number 25, Amer. Math. Soc., Providence, Rhode Island.

[25] O. Hanner, Some theorems on absolute neighborhood retracts, Arkiv for Matematik 1 (1951), 389–408.

[26] ______, Retraction and extension of mappings of metric and non–metric spaces, Arkiv for Matematik 2 (1952–1954), 315–360.

[27] G.S. Jones, The existence of periodic solutions of $f'(x) = -\alpha f(x-1)[1+f(x)]$, J. Math Anal. Appl. 4 (1962), 440–469.

[28] ______, Periodic motions in Banach space and applications to functional differential equations, Contributions to Differential Equations 3 (1964), 75–106.

[29] M.A. Krasnosel'skii and P.P. Zabreiko, Iterations of operators and the fixed point index, Soviet Math. Doklady 12 (1971), 294–298.

[30] C. Kuratowski, Sur les espaces complets, Fund. Math. 15 (1930), 301–309.

[31] J. Leray and J. Schauder, Topologie et équations fonctionnelles, Ann. Sci. Ecole Norm. Sup. 51 (1934), 45–78.

[32] J. Leray, Sur les équations et les transformations, J. Math. Pures Appl. 24 (1945), 201–248.

[33] ______, La theorie des points fixes et ses applications en analyse, Proc. International Congress Math., Cambridge, vol. 2, (1950), pp. 202–208.

[34] ______, Theorie des points fixes, indice total et nombre de Lefschetz, Bull. Soc. Math. France 87 (1959), 221–233.

[35] J. Mallet–Paret and R.D. Nussbaum, Global continuation and asymptotic behaviour for periodic solutions of a differential–delay equation, Annali di Mat. Pura Appl. 145 (1986), 33–128.

[36] J. Mallet–Paret and R.D. Nussbaum, Boundary layer phenomena for differential–delay equations with state dependent time lags, I, to appear in Arch. Rat. Mech. and Analysis

[37] L. Nirenberg, Topics in Nonlinear Functional Analysis, N.Y.U. Courant Institute Lecture Notes, New York, 1974.

[38] R.D. Nussbaum, The fixed point index and asymptotic fixed point theorems for k–set–contractions, Bull. Amer. Math. Soc. 75 (1969), 490–495.

[39] ______, The radius of the essential spectrum, Duke Math J. 38 (1970), 473–478.

[40] ______, The fixed point index for local condensing maps, Ann. Mat. Pura Appl. 89 (1971), 217–258.

[41] ______, A geometric approach to the fixed point index,

Pacific J. Math. 39 (1971), 751–766.

[42] ________________, Asymptotic fixed point theory for local condensing maps, Math. Ann. 191 (1971), 181–195.

[43] ________________, Some asymptotic fixed point theorems, Transactions Amer. Math. Soc. 171 (1972), 349–375.

[44] ________________, Periodic solutions of some nonlinear autonomous functional differential equations, Ann. Mat. Pura Appl. 101 (1974), 263–306.

[45] ________________, Periodic solutions of some nonlinear autonomous functional differential equations, II, J. Differential Equations 14 (1973), 360–394.

[46] ________________, A global bifurcation theorem with applications to functional differential equations, J. Functional Analysis 19 (1975), 319–338.

[47] ________________, The range of periods of periodic solutions of $x'(t) = -\alpha f(x(t-1))$, J. Math. Anal. Appl. 58 (1977), 280–292.

[48] ________________, Periodic solutions of nonlinear, autonomous functional differential equations, in Functional Differential Equations and Approximation of Fixed Points, H.–O. Peitgen and H.–O. Walther, eds., Springer Verlag Lecture Notes in Mathematics, vol. 730, Springer–Verlag, New York, 1979, pp. 283–325.

[49] ________________, Generalizing the fixed point index, Math. Ann. 228 (1977), 259–278.

[50] ________________, A periodicity threshold theorem for some nonlinear integral equations, SIAM J. Math. Anal. 9 (1978), 356–376.

[51] ________________, Periodic solutions of some nonlinear integral equations, in Dynamcal Systems, A. Bednarek and L. Cesari, eds., Academic Press, New York, 1977, pp. 221–251.

[52] ________________, The Fixed Point Index and Some Applications, volume 94 in the collection, "Séminaire de Mathematiques Supérieures", Les Presses de l'Université Montreal, C.P. 6128, succ. "A", Montreal, Québec, Canada H3C3J7, 1985.

[53] ________________, Eigenvectors of nonlinear positive operators and the linear Krein–Rutman theorem, in Fixed Point Theory, E. Fadell and G. Fournier, eds., Springer Lecture Notes in Mathematics, vol. 886, Spring–Verlag, New York, 1981, pp. 309–331.

[54] H.–O. Peitgen, On the Lefschetz number for iterates of continuous mappings, Proc. Amer. Math. Soc. 54 (1976), 441–444.

[55] P. Rabinowitz, Some global results for nonlinear eigenvalue problems, J. Functional Analysis 7 (1971), 487–513.

[56] ________________, Some aspects of nonlinear eigenvalue problems, Rocky Mountain Math. J. 3 (1973), 161–202.

[57] B.N. Sadovskii, Limit compact and condensing operators, Russian Math. Surveys 27 (1972), 85–155.

[58] H. Steinlein, Ein Satz über den Leray–Schauderschen Abbildungsgrad, Math. Z. 120 (1972), 176–208.

[59] ________________, Über die verallgemeinerten Fixpunktindizes von Iterierten verdichtender Abbildungen, Manuscripta Math. 8 (1972), 251–266.

[60] R.B. Thompson, A unified approach to local and global fixed point indices, Advances in Math. 3 (1969), 1–72.

[61] A.J. Tromba, The beer barrel theorem, in Functional Differential Equations and Approximation of Fixed Points, H.–O. Peitgen and H.–O. Walther, eds., Springer Verlag Lecture Notes in Mathematics, vol. 730, pp. 484–488.

[62] ________________, A general asymptotic fixed point theorem, J. für die Reine u. Angew. Math. 332 (1982), 118–125.

C.I.M.E. Session on "Topological methods in the theory of ordinary differential equations in finite and infinite dimensions"

Montecatini Terme, June 24-July 2, 1991

LIST OF PARTICIPANTS

W. ALLEGRETTO, Dept. of Math., Univ. of Alberta, Alberta T6G 2G1

H. AMANN, Math. Inst., Univ. Zürich, Rämistrasse 74, CH-8001 Zürich

G. ANICHINI, Dip. Mat., Univ. Modena, v. Campi 213/b, 41100 Modena

A.K. BEN NAOUM, Dép. Math., UCL, B-1348 Louvain-la-Neuve

P. BILER, Math. Inst., Univ. Wroclaw, Pl. Grunwaldzki 2/4, 50-384 Wroclaw

F. BUCCI, Dip. Mat., Univ. Modena, v. Campi 213/b, 41100 Modena

A. BURMISTROVA, pr. Lenina 77, kv. 5, 454080 Chelyabinsk

G. BUSONI, Dip. Mat., Univ. Firenze, Viale Morgagni 67/A, 50134 Firenze

S. CALAFIORE, Dip. Mat., Univ. Firenze, Viale Morgagni 67/A, 50134 Firenze

M. CALAHORRANO, ICTPh. Math. Group, P.O.Box 586, 34100 Trieste

A.M. CANDELA, Dip. Mat., Univ. Pisa, v. Buonarroti 2, 56127 Pisa

A.M. CANINO, Dip. Mat., Univ. Calabria, 87036 Arcavacata di Rende (Cosenza)

A. CAPIETTO, Dip. Mat., Univ. Torino, v. Carlo Alberto 10, 10123 Torino

A. CARBONE, Dip. Mat., Univ. Calabria, 87036 Arcavacata di Rende (Cosenza)

M. CECCHI, Dip. Mat., Univ. Siena, v. del Capitano 15, 53100 Siena

R. CHIAPPINELLI, Dip. Mat., Univ. Calabria, 87036 Arcavacata di Rende (Cosenza)

G. CICOGNA, Dip. Fisica, Univ. Pisa, Piazza Torricelli 2, 56126 Pisa

K. CIESIELSKI, Math. Inst., Jagellonian Univ. Reymonta 4, 30-050 Krakow

F.S. DE BLASI, Dip. Mat., Univ. Tor Vergata, v. Fontanile di Carcaricola, 00133 Roma

C. DE COSTER, Inst. Math. pure appl., UCL, B-1348 Louvain-la-Neuve

A. DO NASCIMENTO, Rua Cidade de Setubal, 13, 2°E, Corroios, 2800 Almada, Portugal

Z.R. DZEDZEJ, Dept. Math., Gdansk Univ., ul. Wita Stwosza 57, 80-452 Gdansk

J. ESCHER, Math. Inst., Univ. Zürich, Rämistrasse 74, CH-8001 Zürich

A. FONDA, SISSA, Strada Costiera 11, 3404 Trieste

W. FRATTAROLO, Ist. Mat., Univ. Pisa, v. Bonanno Pisano 25/B, 56100 Pisa

M. GIRARDI, Dip. Mat., Univ. L'Aquila, 67100 L'Aquila

P. GROSS, Math. Inst., T.U. Denmark, Building 303, D-2800 Lyngby

A. HANSBO, Dept. Math., Chalmers U. Techn., Göteborg

M. HENRARD, Inst. Math. pure appl., UCL, B-1348 Louvain-la-Neuve

J. JODEL, Inst. Math., Univ. Gdansk, ul. Wita Stwosza 57, 80-952 Gdansk

W. KRYSZEWSKI, Math. Inst., Univ. München, Theresienstr. 39, 8000 München

C. LIM, 40 Church St., Tarryton, NY 10591, USA

T. LINDSTROM, Dept. Math., Lulea Univ., S-95187 Lulea

G. LOPEZ, Dep. Anal. Mat., Univ. Sevilla, Apart. 1160, 41080 Sevilla

D. LUPO, Dip. Sc. Mat., Univ. Trieste, P.le Europa 1, 34100 Trieste

L. MALAGUTI, Dep. Mat., Univ. Modena, v. Campi 213/b, 41100 Modena

C. MARCHIO', Dip. Mat., Univ. Siena, v. del Capitano 15, 53100 Siena

A. MARGHERI, Dip. Mat., Univ. Firenze, Viale Morgagni 67/A, 50134 Firenze

M. MARINI, Dip. Elettrot. Elettr., Fac. Ing., v. S. Marta 3, 50139 Firenze

A. MARZOCCHI, Dip. Mat., Univ. Cattolica S. Cuore, 25121 Brescia

I. MASSABO, Dip. Mat., Univ. Calabria, 87036 Arcavacata di Rende (Cosenza)

M. MATZEU, Dip. Mat., Univ. Tor Vergata, v. Fontanile di Carcaricola, 00133 Roma

C. MICHAUX, Space Sc. Div., Jet Propulsion Lab., Pasadena, CAL 90109

A.M. MICHELETTI, Ist. Mat., Fac. Ing., Univ. Pisa, 56100 Pisa

M. MOLINA MEYER, Fac. Ciencias, Univ. Auton. Madrid, Cantoblanco, 28049 Madrid

M. MROZEK, Kat. Informatyki, Uniw. Jagiellonski, ul. Kopernika 27, PL 31-501 Krakow

J. MYJAK, Dip. Mat., Fac. Scienze, Univ. L'Aquila, v. Roma 33, 67100 L'Aquila

P. NISTRI, Dip. Sist. Info., Fac. Ing., v. S. Marta 3, 50139 Firenze

R. NUGARI, Dip. Mat., Univ. Calabria, 87036 Arcavacata di Rende (Cosenza)

P. OMARI, Dip. Sc. Mat., Univ. Trieste, P.le Europa 1, 34127 Trieste

J. PEJSACHOWICZ, Dip. Mat., Polit. Torino, Corso Duca degli Abruzzi 24, 10129 Torino

M.P. PERA, Dip. Mat. Appl., Fac. Ing., v. S. Marta 3, 50139 Firenze

A. PISTOIA, Dip. Mat., Univ. Pisa, v. Buonarroti 2, 56127 Pisa

B. PRZERADZKI, Inst. Mat., Univ. Lodz, ul. Banacha 22, 90-238 Lodz

A.PUGLIESE, Dip. Mat., Univ. Trento, 38050 Povo (Trento)

E. ROSSET, Dip. Sc. Mat., Univ. Trieste, P.le Europa 1, 34127 Trieste

D. ROUX, Dip. Mat., Univ. Milano, v. C. Saldini 50, 20133 Milano

P. SANTORO, Dip. Mat. Appl., Fac. Ing., v. S. Marta 3, 50139 Firenze

G. SAVARE', IAN-CNR, Palazzo Univ., Corso Carlo Alberto 5, 27100 Pavia

I. SCARASCIA, v. Martiri di Filetto 1, 67100 L'Aquila

G. SIMONETT, Math. Inst., Univ. Zürich, Rämistrasse 74, CH-8001 Zürich

A. SLAVOVA, Dept. Math., Techn. Univ., 7000 Russe, Bulgaria

R. SRZEDNICKI, Inst. Mat., Univ. Jagell., ul. Reymonta 4, 30-059 Krakow

H. STEINLEIN, Math. Inst., Univ. München, Theresienstr. 39, 8000-München 2

S. TERRACINI, Dip. Mat., Pol. Milano, P.za L. da Vinci 32, 20133 Milano

Gab. VILLARI, Dip. Mat., Univ. Firenze, V.le Morgagni 67/A, 50134 Firenze

M. VILLARINI, Dip. Mat., Univ. Firenze, V.le Morgagni 67/A, 50134 Firenze

G. WEILL, Dept. Math., Polyt. Univ., 333 Jay Street, Brooklyn, NY 11201

F. ZANOLIN, Dip. Mat., Univ. Udine, v. Zanon 6, 33100 Udine

FONDAZIONE C.I.M.E.
CENTRO INTERNAZIONALE MATEMATICO ESTIVO
INTERNATIONAL MATHEMATICAL SUMMER CENTER

"Arithmetic Algebraic Geometry"

is the subject of the Second 1991 C.I.M.E. Session.

The Session, sponsored by the Consiglio Nazionale delle Ricerche and by the Ministero dell'Università e della Ricerca Scientifica e Tecnologica, will take place under the scientific direction of Prof. Edoardo BALLICO (Università di Trento) at Villa Madruzzo (Cognola di Trento) **from June 24 to July 2, 1991.**

Courses

a) **Torsion Algebraic Cycles and Algebraic K-Theory.** (6 lectures in English).
Prof. Jean-Louis COLLIOT-THELENE (CNRS, Université de Paris-Sud, Orsay).

Outline

The Bloch/Ogus theory.

Some key facts from Quillen's 1973 K-theory paper.

Some more recent developments: known cases of the Gersten conjecture.

Etale cohomology of varieties defined over arithmetic fields: a summary.

Summary of some results of Merkur'ev and Suslin. Applications to K-cohomology.

Algebraically closed ground fields: Roitman's theorem.

Finite ground fields: codimension two cycles; the unramified class field theory of Kato and Saito.

Arbitrary ground fields: cycle class maps for torsion cycles via Galois action over an algebraic closure.

Varieties over p-adic or number fields: finiteness results for torsion codimension two cycles, via the Galois methed.

Varieties over p-adic or number fields: finiteness results for torsion codimension two cycles, via the localization method.

Local-to-global conjectures for the torsion in the codimension two Chow groups of varieties defined over number fields.

References

S. Bloch, Torsion algebraic cycles and a theorem of Roitman, Comp. Math. 39 (1979), 107-127.

S. Bloch, Lectures on algebraic cycles, Duke Univ. Math. Ser. 4, Durham, 1980.

S. Bloch, Torsion algebraic cycles, K2 and Brauer groups of function fields, in Springer L.N.M. 844 (1981)

S. Bloch, On the Chow groups of certain rational surfaces, Ann. Ec. Norm. Sup. (1981), 41-59.

S. Bloch, Algebraic K-theory and class field theory for arithmetic surfaces, Ann. of Math. 114 (1981), 229-266.

S. Bloch and A. Ogus, Gersten's conjecture and the homology of schemes, Ann. Sc. Ec. Norm. Sup. 4 (1974), 181-202.

S. Bloch, A. Kas and D. Lieberman, Zero-cycles on surfaces with pg = 0, Comp. Math. 33 (1976), 135-145.

S. Bloch and V. Srinivas, Remarks on correspondences and algebraic cycles, Amer. J. Math. 105 (1983), 1235-1253.

J.-L. Colliot-Thélène, Hilbert's theorem 90 for K2, with applications to the Chow groups of rational surfaces, Inventiones Math. 71 (1983), 1-20.

J.-L. Colliot-Thélène and W. Raskind, K2-cohomology and the second Chow group, Math. Annalen 270 (1985), 165-199.

J.-L. Colliot-Thélène and W. Raskind, On the reciprocity law for surfaces over a finite field, J. of the Faculty of Science, University of Tokio, 33 (1986), 282-294.

J.-L. Colliot-Thélène and W. Raskind, Groupe de Chow de codimension deux des variétés définies sur un corps de nombres: un théorème de finitude pour la torsion, to appear in Inventiones Math.
J.-L. Colliot-Thélène et J.-J. Sansuc, On the Chow groups of certain rational surfaces: a sequel to a paper of S. Bloch, Duke Math. J. 48 (1981), 421-447.
J.-L. Colliot-Thélène, J.-J. Sansuc et C. Soulé, Torsion dans le groupe de Chow de codimension deux, Duke Math. J. 48 (1983), 763-801.
K. Coombes, The arithmetic of zero cycles on surfaces with geometric genus zero and irregularity zero, preprint.
M. Gros, O-cycles de degré zéro sur les surfaces fibrées en coniques, J. reine und ang. Math. 373 (1987), 166-184.
U. Jannsen, On the l-adic cohomology of varieties over number fields and its Galois cohomology, in: Galois groups over Q, MSRI Publications.
K. Kato and S. Saito, Unramified class field theory of arithmetical surfaces, Ann. of Math. 118 (1983), 241-275.
A.S. Merkur'ev and A.A. Suslin, K-cohomology of Severi-Brauer varieties and norm residue homomorphism, Izv. Akad. Nauk SSSR 46 (1982), 1011-1046 - Math. USSR Izv. 21 (1983), 307-341.
D. Quillen, Higher algebraic K-theory I, in Springer L.N.M. 341.
W. Raskind, Algebraic K-theory, étale cohomology and torsion algebraic cycles, Contemporary Mathematics 83 (1989), 311-341.
W. Raskind, Torsion algebraic cycles on varieties over local fields, in Algebraic K-theory: Connection with geometry and varieties over local fields, in Algebraic K-theory: Connection with geometry and topology, 343-388, (1989), Kluwer Academic Publishers.
C. Schoen, Some examples of torsion in the Griffiths Group, preprint.
C. Schoen, On the computation of the cycle class map for nullhomologous cycles over the algebraic closure of a finite field, preprint.
S. Saito, A. conjecture of Bloch and Brauer groups of surfaces over p-adic fields, preprint, May 1990.
S. Saito, Cycle map on torsion algebraic cycles of codimension two, preprint, October 1990.
P. Salberger, Zero-cycles on rational surfaces over number fields, Inventiones Math. 91 (1988), 505-524.
A.A. Suslin, Torsion in K2 of fields, Journal of K-Theory, 1 (1987), 5-29.

b) **Values of L-functions, Explicit Reciprocity Laws, and p-adic Hodge Theory.** (6 lectures in English).
Prof. Kazuya KATO (University of Tokyo).

Outline

The main theme is the p-adic property of special values of Hasse-Weil L-functions of algebraic varieties over number fields. By using the p-adic Hodge theory developed by J. Tate, J.M. Fontaine, W. Messing, G. Faltings, and other people, we generalize the classical explicit reciprocity law in local class field theory to more general varieties, and relate it to values of L-functions. Related conjectures on Iwasawa theory of varieties over number fields are proposed and discussed.

References

S. Bloch and K. Kato, L-functions and Tamagawa numbers of motives, in the Grothendieck Festschrift, Vol. I, Birkhäuser 1990, 333-400.
G. Faltings, Crystalline cohomology and p-adic Galois representation, in Algebraic Analysis, Geometry, and Number Theory, The Johns Hopkins Univ. Press, 1989, 25-80.
J.M. Fontaine and W. Messing, p-adic periods and p-adic étale cohomology, in Current Trends in Arithmetical Algebraic Geometry, Contemporary Math. 67, 1987, 179-207.

c) **Arakelov Geometry.** (6 lectures in English).
Prof. Christophe SOULE' (IHES).

Outline

This course will present an overview of joint work with Henri Gillet, extending to arbitrary dimensions the Arakelov theory of arithmetic surfaces. The main result is a Riemann-Roch type theorem for hermitian bundles on arithmetic varieties (generalizing previous work of Faltings and Deligne). Its proof uses recent analytical work of Bismut and others on the determinant of Laplace operators. In this survey we shall also try to indicate the possible links between this geometry and different aspects of number theory.

References

Some familiarity with the basic notions of the algebraic geometry of schemes (Hartshorne's book, Chapters 1 and 2) and complex geometry (Griffiths and Harris, chapters 0, 1, 3) will be useful.

Original papers:

C. Soulé, H. Gillet: Intersection theory using Adams operations, Inventiones Math. 90 (1987), 243-277.

C. Soulé, H. Gillet, Direct images of Hermitian holomorphic bundles, Bull. AMS 15 (1986), 209-212.

J.-M. Bismut, H. Gillet, C. Soulé, Analytic torsion and holomorphic determinant bundles I: Bott-Chern forms and analytic torsion, Comm. in math. Physics 115 (1988), 49-78.

J.-M. Bismut, H. Gillet, C. Soulé, Analytic torsion and holomorphic determinant bundles II: Direct images and Bott-Chern forms, Comm. in Math. Physics 115 (1988), 79-126.

J.-M. Bismut, H. Gillet, C. Soulé, Analytic torsion and holomorphic determinant bundles III: Quillen metrics on holomorphic determinants, Comm. Math. Physics 115 (1988), 301-351.

H. Gillet, C. Soulé, Intersection on arithmetic varieties, Publications IHES, à paraître.

H. Gillet, C. Soulé, Characteristic classes for algebraic vector bundles with hermitian metric, Annals of Math. 131 (1990), 163-203.

H. Gillet, C. Soulé, Analytic torsion and the arithmetic Todd genus, Topology, à paraître.

H. Gillet, C. Soulé, Arithmetic amplitude, note aux CRAS, 307, 1988, 887-890.

H. Gillet, C. Soulé, Un théorème de Riemann-Roch arithmétique, CRAS Paris, 309, 1989, 929-932.

J.-M. Bismut, H. Gillet, C. Soulé, Bott-Chern currents and complex immersions, Duke Math. J. 60 (1990), 255-284.

J.-M. Bismut, H. Gillet, C. Soulé, Complex immersions and Arakelov geometry, Grothendieck Festschrift, Birkhäuser, 1990.

Survey papers:

C. Soulé, Théorie d'Arakelov et théorie de Nevanlinna, Astérisque 183 (1990), 127-135.

C. Soulé, Géométrie d'Arakelov et théorie des nombres transcendants, Astérisque, â paraître.

C. Soulé, Géométrie d'Arakelov des surfaces arithmétiques, exposé au séminaire Bourbaki, Juin 1989, 15 p.

Especially:

J.B. Bosl, Théorie de l'intersection et théorème de Riemann-Roch arithmétiques, Séminaire Bourbaki, Novembre 1990, 37 p.

d) **Applications of Arithmetic Geometry to Diophantine Approximations.** (6 lectures in English).
Prof. Paul VOJTA (University of California, Berkeley).

Outline

1. Thue's method
2. Mordell's conjecture over function fields (Manin's theorem)
3. A proof for number field
4. Enhancements due to Faltings
5. Enhancements due to Bombieri.

References

W.M. Schmidt, Application of Thue's method in various branches of number theory. Proceedings of the International Congress of Mathematicias, R.P. James, ed., Canadian Mathematical Congress, Vancouver, 1975, pp. 177-186.

P. Vojta, Mordell's conjecture over function fields, Invent. Math. 98 (1989), 115-138.

P. Vojta, Siegel's theorem in the compact case, Ann. Math. (to appear).

G. Faltings, Diophantine approximation on abelian varieties, Ann. Math. (to appear).

E. Bombieri, The Mordell conjecture revisited. (to appear)

FONDAZIONE C.I.M.E.
CENTRO INTERNAZIONALE MATEMATICO ESTIVO
INTERNATIONAL MATHEMATICAL SUMMER CENTER

"Transition to Chaos in Classical and Quantum Mechanics"

is the subject of the Third 1991 C.I.M.E. Session.

The Session, sponsored by the Consiglio Nazionale delle Ricerche and by the Ministero dell'Università e della Ricerca Scientifica e Tecnologica, will take place under the scientific direction of Prof. Sandro GRAFFI (Università di Bologna, Dipartimento di Matematica, Piazza di Porta S. Donato, 40126 Bologna, E-mail: MK7BOG73 at ICINECA; graffi at dm.unibo.it.) at Villa "La Querceta", Montecatini (Pistoia), **from July 6 to July 13, 1991.**

Courses

a) **Non Commutative Method in Semi Classical Analysis.** (8 lectures in English).
Prof. Jean BELLISSARD (Wissenschaftskolleg zu Berlin).

Outline

1) Quantum phase space ad groupoid
2) Perturbation theory: Birkhoff versus Rayleigh Schrödinger in the semi classical limit
3) Semi classical expansion and tunneling effects
4) Analysis of resonances: KAM theory and quantum localization
5) Spectrum in the classically chaotic region

References

1. J. Bellissard, C* algebras in solid state physics, in "Operator algebras and applications", D.E. Evans and M. Takesuki Eds., Cambridge Univ. Press 1988.
2. J. Bellissard, M. Vittot, Heisenberg's picture and non commutative geometry of the semi classical limit in quantum mechanics, Ann. Inst. H. Poincaré 52 (1990), 175-235.
3. J. Bellissard, R. Rammal, An algebraic semi-classical approach to Bloch electrons in a magnetic field, J. de Physique Paris, (March 1990).
4. J. Bellissard, Stability and instability in quantum mechanics, in "Trends and developments in the eighties", Albeverio, Blanchard, Eds, World Sc. Publ., 1985, 1-106.
5. O. Bohigas, M.J. Giannoni, C. Schmit, Spectral properties of the Laplacian and random matrix theory, Lecture notes in physics, vol. 209 (1984), p. 1.

b) **Stochastic Properties of Classical Dynamical Systems.** (8 lectures in English).
Prof. Anatole KATOK (Pennsylvania State University).

Outline

1. Review of stochastic properties of dynamical systems. Ergodicity, mixing, eigenfunctions. Entropy and asymptotic independence. K-property, Bernoulli property.

2. Hyperbolic systems with examples exhibiting stochastic properties and complicate dynamical behaviour.
3. Smooth hyperbolic systems. The Pesin theory. Lyapunov exponents, stable and unstable manifolds, stochastic behaviour. The Pesin entropy formula, local ergodicity and local Bernoulli property.
4. Application of Pesin's theory to specific systems. Invariant cone families and Lyapynov characteristic exponents. Wojtkoski's theorem. Symplectic cones.
5. From local to global ergodicity. Criteria of openness of ergodic components and ergodicity based on existence of virtually strictly invariant family of symplectic cones.
6. Dynamical systems with singularities. Billiards, collisions, elastic balls in a volume and multidimensional billiard. Generalization of the Pesin theory to include billiards.
7. Cone families for various special classes of classical dynamical systems, both smooth and with singularities.
8. Criteria of ergodicity for dynamical systems with singularities.

References

1. Walters, P.: An introduction to ergodic theory, Springer Verlag, 1981.
2. Katok, A.: Dynamical systems with hyperbolic structure, in: Three papers in dynamical systems, AMS Translations (2) Vol. 116, 1981.
3. Pesin, Ya. B.: Characteristic Lyapunov exponents and smooth ergodic theory, Russian Math. Surveys, Vol. 32 (1977), 54-114.
4. Katok, A.: Lyapunov exponents, entropy and periodic points of diffeomorphisms, Publ. Math. IHES, Vol. 51 (1980), 137-173.
5. Wojtkowski, M.: Invariant families of cones and Lyapunov exponents, Ergodic Theory and Dynamical Systems, Vol. 5 (1985), 145-161.

c) **Dynamics of Area Preserving Maps.** (8 lectures in English).
Prof. John N. MATHER (Princeton University).

Outline

1. Hyperbolic and elliptic fixed points, Birkhoff normal form
2. A brief discussion of KAM theory (no proofs)
3. A brief discussion on the "last invariant circle" with reference to the numerical results of Greene, Percival, McKay (no proofs)
4. An outline of Aubry's theory and of its generalization by Barget
5,6. Herman's method and the author's method of destroying invariant circles (with an outline of the proofs)
7,8. Orbits which pass close to a succession of Aubry-Mather sets (with an outline of the proofs)

References

Herman, M.R.: Sur les courbes invariantes par les difféomorphismes de l'anneau. Vol. 1, Astérisque, 103-104.
Bangert, V.: Mather sets for twist maps and geodesics on tori, Dynamics reported, Vol. 1, pp. 1-56.
Salamon, S. And Zehnder, E.: The Kolmogorov-Arnold-Moser theory in configuration space, Comm. Math. Helv., Vol. 64, pp. 84-132.
Mather, J.N.: Destruction of invariant circles, Ergodic Theory and Dynamical Systems, 8, 199-214.
Mather, J.N.: Variational construction of orbits of twist diffeomorphisms, Journal of the American Mathematical Society, to appear.

LIST OF C.I.M.E. SEMINARS **Publisher**

Year	Seminar	Publisher
1954 -	1. Analisi funzionale	C.I.M.E.
	2. Quadratura delle superficie e questioni connesse	"
	3. Equazioni differenziali non lineari	"
1955 -	4. Teorema di Riemann-Roch e questioni connesse	"
	5. Teoria dei numeri	"
	6. Topologia	"
	7. Teorie non linearizzate in elasticità, idrodinamica,aerodinamica	"
	8. Geometria proiettivo-differenziale	"
1956 -	9. Equazioni alle derivate parziali a caratteristiche reali	"
	10. Propagazione delle onde elettromagnetiche	"
	11. Teoria della funzioni di più variabili complesse e delle funzioni automorfe	"
1957 -	12. Geometria aritmetica e algebrica (2 vol.)	"
	13. Integrali singolari e questioni connesse	"
	14. Teoria della turbolenza (2 vol.)	"
1958 -	15. Vedute e problemi attuali in relatività generale	"
	16. Problemi di geometria differenziale in grande	"
	17. Il principio di minimo e le sue applicazioni alle equazioni funzionali	"
1959 -	18. Induzione e statistica	"
	19. Teoria algebrica dei meccanismi automatici (2 vol.)	"
	20. Gruppi, anelli di Lie e teoria della coomologia	"
1960 -	21. Sistemi dinamici e teoremi ergodici	"
	22. Forme differenziali e loro integrali	"
1961 -	23. Geometria del calcolo delle variazioni (2 vol.)	"
	24. Teoria delle distribuzioni	"
	25. Onde superficiali	"
1962 -	26. Topologia differenziale	"
	27. Autovalori e autosoluzioni	"
	28. Magnetofluidodinamica	"

1963 - 29. Equazioni differenziali astratte "
30. Funzioni e varietà complesse "
31. Proprietà di media e teoremi di confronto in Fisica Matematica "

1964 - 32. Relatività generale "
33. Dinamica dei gas rarefatti "
34. Alcune questioni di analisi numerica "
35. Equazioni differenziali non lineari "

1965 - 36. Non-linear continuum theories "
37. Some aspects of ring theory "
38. Mathematical optimization in economics "

1966 - 39. Calculus of variations Ed. Cremonese, Firenze
40. Economia matematica "
41. Classi caratteristiche e questioni connesse "
42. Some aspects of diffusion theory "

1967 - 43. Modern questions of celestial mechanics "
44. Numerical analysis of partial differential equations "
45. Geometry of homogeneous bounded domains "

1968 - 46. Controllability and observability "
47. Pseudo-differential operators "
48. Aspects of mathematical logic "

1969 - 49. Potential theory "
50. Non-linear continuum theories in mechanics and physics and their applications "
51. Questions of algebraic varieties "

1970 - 52. Relativistic fluid dynamics "
53. Theory of group representations and Fourier analysis "
54. Functional equations and inequalities "
55. Problems in non-linear analysis "

1971 - 56. Stereodynamics "
57. Constructive aspects of functional analysis (2 vol.) "
58. Categories and commutative algebra "

1972 -	59. Non-linear mechanics		"
	60. Finite geometric structures and their applications		"
	61. Geometric measure theory and minimal surfaces		"
1973 -	62. Complex analysis		"
	63. New variational techniques in mathematical physics		"
	64. Spectral analysis		"
1974 -	65. Stability problems		"
	66. Singularities of analytic spaces		"
	67. Eigenvalues of non linear problems		"
1975 -	68. Theoretical computer sciences		"
	69. Model theory and applications		"
	70. Differential operators and manifolds		"
1976 -	71. Statistical Mechanics		Ed Liguori, Napoli
	72. Hyperbolicity		"
	73. Differential topology		"
1977 -	74. Materials with memory		"
	75. Pseudodifferential operators with applications		"
	76. Algebraic surfaces		"
1978 -	77. Stochastic differential equations		"
	78. Dynamical systems		Ed Liguori, Napoli and Birhäuser Verlag
1979 -	79. Recursion theory and computational complexity		"
	80. Mathematics of biology		"
1980 -	81. Wave propagation		"
	82. Harmonic analysis and group representations		"
	83. Matroid theory and its applications		"
1981 -	84. Kinetic Theories and the Boltzmann Equation	(LNM 1048)	Springer-Verlag
	85. Algebraic Threefolds	(LNM 947)	"
	86. Nonlinear Filtering and Stochastic Control	(LNM 972)	"
1982 -	87. Invariant Theory	(LNM 996)	"
	88. Thermodynamics and Constitutive Equations	(LN Physics 228)	"
	89. Fluid Dynamics	(LNM 1047)	"

1983 -	90.	Complete Intersections	(LNM 1092)	"
	91.	Bifurcation Theory and Applications	(LNM 1057)	"
	92.	Numerical Methods in Fluid Dynamics	(LNM 1127)	"
1984 -	93.	Harmonic Mappings and Minimal Immersions	(LNM 1161)	"
	94.	Schrödinger Operators	(LNM 1159)	"
	95.	Buildings and the Geometry of Diagrams	(LNM 1181)	"
1985 -	96.	Probability and Analysis	(LNM 1206)	"
	97.	Some Problems in Nonlinear Diffusion	(LNM 1224)	"
	98.	Theory of Moduli	(LNM 1337)	"
1986 -	99.	Inverse Problems	(LNM 1225)	"
	100.	Mathematical Economics	(LNM 1330)	"
	101.	Combinatorial Optimization	(LNM 1403)	"
1987 -	102.	Relativistic Fluid Dynamics	(LNM 1385)	"
	103.	Topics in Calculus of Variations	(LNM 1365)	"
1988 -	104.	Logic and Computer Science	(LNM 1429)	"
	105.	Global Geometry and Mathematical Physics	(LNM 1451)	"
1989 -	106.	Methods of nonconvex analysis	(LNM 1446)	"
	107.	Microlocal Analysis and Applications	(LNM 1495)	"
1990 -	108.	Geoemtric Topology: Recent Developments	(LNM 1504)	"
	109.	H_∞ Control Theory	(LNM 1496)	"
	110.	Mathematical Modelling of Industrical Processes	(LNM 1521)	"
1991 -	111.	Topological Methods for Ordinary Differential Equations	(LNM 1537)	
	112.	Arithmetic Algebraic Geometry	to appear	"
	113.	Transition to Chaos in Classical and Quantum Mechanics	to appear	"
1992 -	114.	Dirichlet Forms	to appear	
	115.	D-Modules and Representation Theory	to appear	
	116.	Nonequilibrium Problems in Many-Particle Systems	to appear	

Printing: Druckhaus Beltz, Hemsbach
Binding: Buchbinderei Schäffer, Grünstadt

Vol. 1440: P. Latiolais (Ed.), Topology and Combinatorial Groups Theory. Seminar, 1985–1988. VI, 207 pages. 1990.

Vol. 1441: M. Coornaert, T. Delzant, A. Papadopoulos. Géométrie et théorie des groupes. X, 165 pages. 1990.

Vol. 1442: L. Accardi, M. von Waldenfels (Eds.), Quantum Probability and Applications V. Proceedings, 1988. VI, 413 pages. 1990.

Vol. 1443: K.H. Dovermann, R. Schultz, Equivariant Surgery Theories and Their Periodicity Properties. VI, 227 pages. 1990.

Vol. 1444: H. Korezlioglu, A.S. Ustunel (Eds.), Stochastic Analysis and Related Topics VI. Proceedings, 1988. V, 268 pages. 1990.

Vol. 1445: F. Schulz, Regularity Theory for Quasilinear Elliptic Systems and – Monge Ampère Equations in Two Dimensions. XV, 123 pages. 1990.

Vol. 1446: Methods of Nonconvex Analysis. Seminar, 1989. Editor: A. Cellina. V, 206 pages. 1990.

Vol. 1447: J.-G. Labesse, J. Schwermer (Eds), Cohomology of Arithmetic Groups and Automorphic Forms. Proceedings, 1989. V, 358 pages. 1990.

Vol. 1448: S.K. Jain, S.R. López-Permouth (Eds.), Non-Commutative Ring Theory. Proceedings, 1989. V, 166 pages. 1990.

Vol. 1449: W. Odyniec, G. Lewicki, Minimal Projections in Banach Spaces. VIII, 168 pages. 1990.

Vol. 1450: H. Fujita, T. Ikebe, S.T. Kuroda (Eds.), Functional-Analytic Methods for Partial Differential Equations. Proceedings, 1989. VII, 252 pages. 1990.

Vol. 1451: L. Alvarez-Gaumé, E. Arbarello, C. De Concini, N.J. Hitchin, Global Geometry and Mathematical Physics. Montecatini Terme 1988. Seminar. Editors: M. Francaviglia, F. Gherardelli. IX, 197 pages. 1990.

Vol. 1452: E. Hlawka, R.F. Tichy (Eds.), Number-Theoretic Analysis. Seminar, 1988–89. V, 220 pages. 1990.

Vol. 1453: Yu.G. Borisovich, Yu.E. Gliklikh (Eds.), Global Analysis – Studies and Applications IV. V, 320 pages. 1990.

Vol. 1454: F. Baldassari, S. Bosch, B. Dwork (Eds.), p-adic Analysis. Proceedings, 1989. V, 382 pages. 1990.

Vol. 1455: J.-P. Françoise, R. Roussarie (Eds.), Bifurcations of Planar Vector Fields. Proceedings, 1989. VI, 396 pages. 1990.

Vol. 1456: L.G. Kovács (Ed.), Groups – Canberra 1989. Proceedings. XII, 198 pages. 1990.

Vol. 1457: O. Axelsson, L.Yu. Kolotilina (Eds.), Preconditioned Conjugate Gradient Methods. Proceedings, 1989. V, 196 pages. 1990.

Vol. 1458: R. Schaaf, Global Solution Branches of Two Point Boundary Value Problems. XIX, 141 pages. 1990.

Vol. 1459: D. Tiba, Optimal Control of Nonsmooth Distributed Parameter Systems. VII, 159 pages. 1990.

Vol. 1460: G. Toscani, V. Boffi, S. Rionero (Eds.), Mathematical Aspects of Fluid Plasma Dynamics. Proceedings, 1988. V, 221 pages. 1991.

Vol. 1461: R. Gorenflo, S. Vessella, Abel Integral Equations. VII, 215 pages. 1991.

Vol. 1462: D. Mond, J. Montaldi (Eds.), Singularity Theory and its Applications. Warwick 1989, Part I. VIII, 405 pages. 1991.

Vol. 1463: R. Roberts, I. Stewart (Eds.), Singularity Theory and its Applications. Warwick 1989, Part II. VIII, 322 pages. 1991.

Vol. 1464: D. L. Burkholder, E. Pardoux, A. Sznitman, Ecole d'Eté de Probabilités de Saint- Flour XIX-1989. Editor: P. L. Hennequin. VI, 256 pages. 1991.

Vol. 1465: G. David, Wavelets and Singular Integrals on Curves and Surfaces. X, 107 pages. 1991.

Vol. 1466: W. Banaszczyk, Additive Subgroups of Topological Vector Spaces. VII, 178 pages. 1991.

Vol. 1467: W. M. Schmidt, Diophantine Approximations and Diophantine Equations. VIII, 217 pages. 1991.

Vol. 1468: J. Noguchi, T. Ohsawa (Eds.), Prospects in Complex Geometry. Proceedings, 1989. VII, 421 pages. 1991.

Vol. 1469: J. Lindenstrauss, V. D. Milman (Eds.), Geometric Aspects of Functional Analysis. Seminar 1989-90. XI, 191 pages. 1991.

Vol. 1470: E. Odell, H. Rosenthal (Eds.), Functional Analysis. Proceedings, 1987-89. VII, 199 pages. 1991.

Vol. 1471: A. A. Panchishkin, Non-Archimedean L-Functions of Siegel and Hilbert Modular Forms. VII, 157 pages. 1991.

Vol. 1472: T. T. Nielsen, Bose Algebras: The Complex and Real Wave Representations. V, 132 pages. 1991.

Vol. 1473: Y. Hino, S. Murakami, T. Naito, Functional Differential Equations with Infinite Delay. X, 317 pages. 1991.

Vol. 1474: S. Jackowski, B. Oliver, K. Pawałowski (Eds.), Algebraic Topology, Poznań 1989. Proceedings. VIII, 397 pages. 1991.

Vol. 1475: S. Busenberg, M. Martelli (Eds.), Delay Differential Equations and Dynamical Systems. Proceedings, 1990. VIII, 249 pages. 1991.

Vol. 1476: M. Bekkali, Topics in Set Theory. VII, 120 pages. 1991.

Vol. 1477: R. Jajte, Strong Limit Theorems in Noncommutative L_2-Spaces. X, 113 pages. 1991.

Vol. 1478: M.-P. Malliavin (Ed.), Topics in Invariant Theory. Seminar 1989-1990. VI, 272 pages. 1991.

Vol. 1479: S. Bloch, I. Dolgachev, W. Fulton (Eds.), Algebraic Geometry. Proceedings, 1989. VII, 300 pages. 1991.

Vol. 1480: F. Dumortier, R. Roussarie, J. Sotomayor, H. Żoładek, Bifurcations of Planar Vector Fields: Nilpotent Singularities and Abelian Integrals. VIII, 226 pages. 1991.

Vol. 1481: D. Ferus, U. Pinkall, U. Simon, B. Wegner (Eds.), Global Differential Geometry and Global Analysis. Proceedings, 1991. VIII, 283 pages. 1991.

Vol. 1482: J. Chabrowski, The Dirichlet Problem with L^2-Boundary Data for Elliptic Linear Equations. VI, 173 pages. 1991.

Vol. 1483: E. Reithmeier, Periodic Solutions of Nonlinear Dynamical Systems. VI, 171 pages. 1991.

Vol. 1484: H. Delfs, Homology of Locally Semialgebraic Spaces. IX, 136 pages. 1991.

Vol. 1485: J. Azéma, P. A. Meyer, M. Yor (Eds.), Séminaire de Probabilités XXV. VIII, 440 pages. 1991.

Vol. 1486: L. Arnold, H. Crauel, J.-P. Eckmann (Eds.), Lyapunov Exponents. Proceedings, 1990. VIII, 365 pages. 1991.

Vol. 1487: E. Freitag, Singular Modular Forms and Theta Relations. VI, 172 pages. 1991.

Vol. 1488: A. Carboni, M. C. Pedicchio, G. Rosolini (Eds.), Category Theory. Proceedings, 1990. VII, 494 pages. 1991.

Vol. 1489: A. Mielke, Hamiltonian and Lagrangian Flows on Center Manifolds. X, 140 pages. 1991.

Vol. 1490: K. Metsch, Linear Spaces with Few Lines. XIII, 196 pages. 1991.

Vol. 1491: E. Lluis-Puebla, J.-L. Loday, H. Gillet, C. Soulé, V. Snaith, Higher Algebraic K-Theory: an overview. IX, 164 pages. 1992.

Vol. 1492: K. R. Wicks, Fractals and Hyperspaces. VIII, 168 pages. 1991.

Vol. 1493: E. Benoît (Ed.), Dynamic Bifurcations. Proceedings, Luminy 1990. VII, 219 pages. 1991.

Vol. 1494: M.-T. Cheng, X.-W. Zhou, D.-G. Deng (Eds.), Harmonic Analysis. Proceedings, 1988. IX, 226 pages. 1991.

Vol. 1495: J. M. Bony, G. Grubb, L. Hörmander, H. Komatsu, J. Sjöstrand, Microlocal Analysis and Applications. Montecatini Terme, 1989. Editors: L. Cattabriga, L. Rodino. VII, 349 pages. 1991.

Vol. 1496: C. Foias, B. Francis, J. W. Helton, H. Kwakernaak, J. B. Pearson, H_∞-Control Theory. Como, 1990. Editors: E. Mosca, L. Pandolfi. VII, 336 pages. 1991.

Vol. 1497: G. T. Herman, A. K. Louis, F. Natterer (Eds.), Mathematical Methods in Tomography. Proceedings 1990. X, 268 pages. 1991.

Vol. 1498: R. Lang, Spectral Theory of Random Schrödinger Operators. X, 125 pages. 1991.

Vol. 1499: K. Taira, Boundary Value Problems and Markov Processes. IX, 132 pages. 1991.

Vol. 1500: J.-P. Serre, Lie Algebras and Lie Groups. VII, 168 pages. 1992.

Vol. 1501: A. De Masi, E. Presutti, Mathematical Methods for Hydrodynamic Limits. IX, 196 pages. 1991.

Vol. 1502: C. Simpson, Asymptotic Behavior of Monodromy. V, 139 pages. 1991.

Vol. 1503: S. Shokranian, The Selberg-Arthur Trace Formula (Lectures by J. Arthur). VII, 97 pages. 1991.

Vol. 1504: J. Cheeger, M. Gromov, C. Okonek, P. Pansu, Geometric Topology: Recent Developments. Editors: P. de Bartolomeis, F. Tricerri. VII, 197 pages. 1991.

Vol. 1505: K. Kajitani, T. Nishitani, The Hyperbolic Cauchy Problem. VII, 168 pages. 1991.

Vol. 1506: A. Buium, Differential Algebraic Groups of Finite Dimension. XV, 145 pages. 1992.

Vol. 1507: K. Hulek, T. Peternell, M. Schneider, F.-O. Schreyer (Eds.), Complex Algebraic Varieties. Proceedings, 1990. VII, 179 pages. 1992.

Vol. 1508: M. Vuorinen (Ed.), Quasiconformal Space Mappings. A Collection of Surveys 1960-1990. IX, 148 pages. 1992.

Vol. 1509: J. Aguadé, M. Castellet, F. R. Cohen (Eds.), Algebraic Topology - Homotopy and Group Cohomology. Proceedings, 1990. X, 330 pages. 1992.

Vol. 1510: P. P. Kulish (Ed.), Quantum Groups. Proceedings, 1990. XII, 398 pages. 1992.

Vol. 1511: B. S. Yadav, D. Singh (Eds.), Functional Analysis and Operator Theory. Proceedings, 1990. VIII, 223 pages. 1992.

Vol. 1512: L. M. Adleman, M.-D. A. Huang, Primality Testing and Abelian Varieties Over Finite Fields. VII, 142 pages. 1992.

Vol. 1513: L. S. Block, W. A. Coppel, Dynamics in One Dimension. VIII, 249 pages. 1992.

Vol. 1514: U. Krengel, K. Richter, V. Warstat (Eds.), Ergodic Theory and Related Topics III, Proceedings, 1990. VIII, 236 pages. 1992.

Vol. 1515: E. Ballico, F. Catanese, C. Ciliberto (Eds.), Classification of Irregular Varieties. Proceedings, 1990. VII, 149 pages. 1992.

Vol. 1516: R. A. Lorentz, Multivariate Birkhoff Interpolation. IX, 192 pages. 1992.

Vol. 1517: K. Keimel, W. Roth, Ordered Cones and Approximation. VI, 134 pages. 1992.

Vol. 1518: H. Stichtenoth, M. A. Tsfasman (Eds.), Codin Theory and Algebraic Geometry. Proceedings, 1991. VIII, 22 pages. 1992.

Vol. 1519: M. W. Short, The Primitive Soluble Permutatio Groups of Degree less than 256. IX, 145 pages. 1992.

Vol. 1520: Yu. G. Borisovich, Yu. E. Gliklikh (Eds.), Glob Analysis – Studies and Applications V. VII, 284 pages. 1992

Vol. 1521: S. Busenberg, B. Forte, H. K. Kuiken, Mathematic Modelling of Industrial Process. Bari, 1990. Editors: V. Capass A. Fasano. VII, 162 pages. 1992.

Vol. 1522: J.-M. Delort, F. B. I. Transformation. VII, 101 pages 1992.

Vol. 1523: W. Xue, Rings with Morita Duality. X, 168 pages 1992.

Vol. 1524: M. Coste, L. Mahé, M.-F. Roy (Eds.), Real Algebrai Geometry. Proceedings, 1991. VIII, 418 pages. 1992.

Vol. 1525: C. Casacuberta, M. Castellet (Eds.), Mathematica Research Today and Tomorrow. VII, 112 pages. 1992.

Vol. 1526: J. Azéma, P. A. Meyer, M. Yor (Eds.), Séminaire d Probabilités XXVI. X, 633 pages. 1992.

Vol. 1527: M. I. Freidlin, J.-F. Le Gall, Ecole d'Eté d Probabilités de Saint-Flour XX – 1990. Editor: P. L. Hennequin VIII, 244 pages. 1992.

Vol. 1528: G. Isac, Complementarity Problems. VI, 297 pages 1992.

Vol. 1529: J. van Neerven, The Adjoint of a Semigroup of Linea Operators. X, 195 pages. 1992.

Vol. 1530: J. G. Heywood, K. Masuda, R. Rautmann, S. A Solonnikov (Eds.), The Navier-Stokes Equations II – Theor and Numerical Methods. IX, 322 pages. 1992.

Vol. 1531: M. Stoer, Design of Survivable Networks. IV, 20 pages. 1992.

Vol. 1532: J. F. Colombeau, Multiplication of Distributions. X 184 pages. 1992.

Vol. 1533: P. Jipsen, H. Rose, Varieties of Lattices. X, 162 pages 1992.

Vol. 1534: C. Greither, Cyclic Galois Extensions of Commutative Rings. X, 145 pages. 1992.

Vol. 1535: A. B. Evans, Orthomorphism Graphs of Groups. VIII, 114 pages. 1992.

Vol. 1536: M. K. Kwong, A. Zettl, Norm Inequalities for Derivatives and Differences. VII, 150 pages. 1992.

Vol. 1537: P. Fitzpatrick, M. Martelli, J. Mawhin, R. Nussbaum, Topological Methods for Ordinary Differential Equations. Montecatini Terme, 1991. Editors: M. Furi, P. Zecca. VII, 218 pages. 1993.